INTERNATIONAL CENTRE FOR MECHANICAL SCIENCES

COURSES AND LECTURES - No. 84

SÁNDOR CSIBI
T.U. OF BUDAPEST, TELECOM. RES. INST.
BUDAPEST, HUNGARY

STOCHASTIC PROCESSES WITH LEARNING PROPERTIES

SPRINGER-VERLAG WIEN GMBH

Originally published by Springer-Verlag Wien-New York in 1975

ISBN 978-3-211-81337-9 ISBN 978-3-7091-3006-3 (eBook)
DOI 10.1007/978-3-7091-3006-3

This Volume includes the Lecture Notes of two distinct series of talks given by the author at subsequent Summer Courses at the International Centre of Mechanical Sciences, Udine, Italy, within Statistical Learning Processes.

Here we publish these two series of fairly self-supporting but closely interrelated notes as distinct monographs, almost in the same way as the topics have been treated in the course of the aforementioned lectures.

While the title of the monograph "Simple and Compound Processes in Machine Learning" is that of the corresponding CISM-course, former participants may like to know that the monograph entitled "Learning Processes: Finite Step Behavior and Approximations" corresponds to the CISM-course "Stability and Complexity of Learning Processes". This revised title appears within this combined material more appropriate to the author.

The Editorial Board

COURSE I

SIMPLE AND COMPOUND PROCESSES IN ITERATIVE MACHINE LEARNING

PREFACE

In this series of talks we consider a fundamental class of disciplines by which algorithms may be made to learn a set of parameters (a function or some more general subject) which is either taught or, in nonsupervised learning, seeked for.

We present in this way an overview of constraints by which the convergence of various simple iterations may be guaranteed. We are in this respect concerned with various sort of procedures governed by (i) costs (ii) potential function type kernels and (iii) such sequences of kernels the convergence constraints of which is inherently specified in terms of long run behavior.

We prove constraints under which additional (e.g. heuristic) processing steps may be embedded, while retaining convergence, into algorithms of guaranteed learning properties. Rich possibilities are offered in this way to flexibly combine efficient heuristic opening procedures, taking specific actual properties into account, with simple mid and end procedures of guaranteed mathematical properties.

We also derive constraints under which the quantities, computed and stored, may be confined to finite sets, and the complexity and storage capacity kept under appropriate control.

In so doing, we make an account of a

number of results in Probability, Statistics, Control Theory and Machine Learning, including also contributions of the speaker's colleagues, and work done by the present speaker.

It is of course, for practical use fairly immaterial how one arrives at some good learning algorithm, provided sufficient experience has already been gathered within the field of actual interest, concerning learning capabilities and efficiency. At the other end,there are always questions left for actual case studies,even if one has got powerful theorems on convergence properties.However,much time may be saved and much better overview obtained if one may specify general constraints under which the learning capability of an algorithm is assured. This is while the knowledge of such constraints is relevant in solving actual machine learning problems.

Many of the topics, we are concerned with, raise in a peculiar way questions in Probability, Statistics and Control Theory. We do our best to emphasize these aspects in the sequel.

We are concerned in what follows, specifically with a collection of disciplines, and this may mislead those mainly interested in the proportion of questions involved in some actual project. It is therefore appropriate to emphasize that what we present in these lectures is just one analytic tool within the rich set of skills, heuristic techniques, combinatorial and analytic disciplines of interest in solving actual problems in machine learning and automatic pattern recognition.

S. CSIBI

CHAPTER I

BASIC NOTIONS

1.1 Machine Learning

Many of the problems, what one likes to term in Computer Science machine learning, may be formulated as follows:

Assume samples ω_i, $i = 0,1,\ldots$ drawn after another from some set Ω, which is, say, an Euclidean space or some subset of it. Members $\omega \in \Omega$ may directly characterize results of actual measurements, observations, objects or symptoms. However, it is more appropriate to think of an Ω that is the collection of features we have derived from such data by means of some appropriate many-to-one mapping. (Many of the interesting and crucial techniques of pattern recognition are concerned just with such feature extraction procedures. However, in what follows, the actual interpretation of Ω does not make much matter.)

We say to learn with a teacher if, together with each ω_i, also some label ℓ_i is given (by a teacher) and the target of learning is specified according to this in some way illustrated, say, by the following Cases 1 through 3. (We will denote the labelled sample observed at i by $\zeta_i = (\omega_i, \ell_i)$.)

<u>Case 1</u> $\ell_i = \Theta(\omega)$ where $\Theta = \{\Theta(\omega), \omega \in \Omega\}$ is, say, a single and real valued function, known only by the teacher.

This is the case of teaching a function Θ , specifically, by representing the values it takes at ω_i, $i = 0,1,\ldots$ One wants, in this case, to derive at some instant n , from the features ω_i and labels ℓ_i $(i \leq n)$ an estimate for Θ in some meaningful way.

<u>Case 2</u> Assume that, at each instant $i = 0, \pm 1, \ldots,$ one of two hypotheses H_0 and H_1 , may hold, and the samples $\omega \in A \cup B \subset \Omega$ are associated with these admissible hypotheses unambigously, in the sense that

$$\omega \in \begin{cases} A & \text{if} \quad H_0 \quad \text{holds} \\ B & \text{if} \quad H_1 \quad \text{holds} \end{cases} \tag{1.1.1}$$

($A \cup B \subset \Omega$. Within Ω $(A \cup B)$ it is irrelevant which of the hypotheses holds.)

Assume that a teacher declares, together with ω_i $\ell_i = \operatorname{sign} \Theta(\omega_i)$ where

$$\operatorname{sign} \Theta(\omega_i) = \begin{cases} -1 & \text{if} \quad H_0 \quad \text{holds} \\ +1 & \text{if} \quad H_1 \quad \text{holds} \end{cases} \tag{1.1.2}$$

While the discriminating function $\Theta = \{\Theta(\omega), \ \omega \in \Omega\}$ has to be single-valued, it may be, apart from this, defined within the scope of (1.1.2) still in many kind of ways.

This is a typical case of hypothesis testing (or in other words classification, decision) problems, with the important simple property that, for any sample $\omega \in A \cup B$ at most one of the hypotheses H_0 and H_1 may hold; i.e. $\ell = \{\ell(\omega),\ \omega \in A \cup B\}$ is a single valued (unambiguous) function.

In Case 2 we can, obviously, experience only unambiguous teaching, in the sense that $\ell_i = \ell_j$ if $\omega_i = \omega_j$.

(1.1.2) may appear as a cause of undue ambiguity. However, just this ambiguity in Θ offers useful freedom for adopting appropriate real-valued estimates.

What one, in this case, finally wants is to give an estimate on A and B or, what implies this, that of sign Θ, at some instant n; which may be done by proposing in $A \cup B$ a decision function.

<u>Case 3</u> We still assume that at each instant $i = 0, \pm 1, \ldots$ either of the hypotheses H_0 and H_1 holds. Assume that a teacher declares, together with ω_i, $\ell_i = d_i$. Here

$$d_i = \begin{cases} 0 & \text{if} \quad H_0 \quad \text{holds} \\ 1 & \text{if} \quad H_1 \quad \text{holds} \end{cases} \tag{1.1.3}$$

We may, of course, describe in this way also such unambiguous testing problems, for which $A \cup B = \Omega$. However, in contrast to Case 2, the possibility of having $\ell_i \neq \ell_j$ if $\omega_i = \omega_j$ is, by

(1.1.3), not excluded.

Obviously, if such ambiguity may also occur, no subdivision A, B $(A \cup B = \Omega)$ may guarantee the correct declaration of H_0 and H_1. In Case 3 we are to create at some n, from observations ω_i and labels ℓ_i $(i \leq n)$, according to this an appropriate subdivision, or a binary valued decision function $\delta = \{\delta(\omega), \omega \in \Omega\}$. Viz., such one, for which the false decisions $d_i \neq \delta(\omega_i)$ are, as far as possible, eliminated. (Let δ take the values either 0 or 1. Declare, for ω_i, H_i if $\delta(\omega_i) = 1$.)

It is well known how to do this in the case we have a probabilistic model and the underlying probability distribution is, a priori, known. However our problem in the present context is, how to devise such estimates for typical learning models, if there is almost none such a priori knowledge.

Up to now we have confined ourselves to learning with a teacher. However, it is usual also to adopt the term learning outside the scope of such problems. It may happen that together with ω_i $(i = 0, \pm 1, \ldots)$ no label ℓ_i is presented (in the aforementioned sense). However we may have some a priori accepted principle of clustering, by which to sort, at some instant n, the samples $\omega_i, i \leq n$ into M classes, and according to this may propose some meaningful subdivision $\{\Omega_1, \ldots, \Omega_M\}$ of Ω. $(\bigcup_{i=1}^{M} \Omega_i = \Omega$. In some of such problems M is a priori fixed, in others also M is derived in the course of learning.

These sort of problems are usually called either

nonsupervised learning or learning without a teacher. The topic is within, what is called in Statistics, cluster analysis.

Let us illustrate, how a clustering principle looks like, and how may the target of learning be defined in this case.

Case 4 Assume, for the sake of simplicity, that our problem is to subdivide Ω into two parts, A and $B (A \cup B = \Omega$, by means of nonsupervised learning, using an a priori given evaluation rule at each instant $i = 0, \pm 1, \dots$. Many sort of such procedures are well known. We pick here just one of these, which may offer some idea about such problems, without entering into detailed statistical considerations.

Assume that at some instant i we have already defined a subdivision by picking two distinct points a_i and b_i , and then subdividing Ω into A_i and $B_i (A_i \cup B_i = \Omega)$ by the nearest neighbor rule. Viz., let

$$\omega \in \begin{cases} A_i & \text{if} \quad \|\omega - a_i\| \le \|\omega - b_i\|, \\ B_i & \text{otherwise} \end{cases} \tag{1.1.4}$$

($\|.\|$ stands for the Euclidean norm in Ω .)

Let us introduce for checking the appropriateness of the subdivision (A_i, B_i) at instant i , the following cost:

$$\tilde{Y}_i(\omega) = \chi_{A_i}(\omega) \|\omega - a_i\|^2 + \chi_{B_i}(\omega) \|\omega - b_i\|^2, \tag{1.1.5}$$

for all $\omega \in \Omega$. Here χ_{A_i} is the indicator of A_i and χ_{B_i} that of B_i. (This specific way of non-supervised learning is well known from Cypkin and Kelmans (1968). See also Vasil'ev (1969)).

Obviously, this definition associates with any point $\omega \in A_i$ a higher cost if it lies further from the template a_i of the class A_i it has been, at instant i , associated with.

In Case 4 the purpose of learning is defined as a choice of such pair of templates $\Theta = (a, b)$ for which the cost, we may expect for any further sample drawn, is in some average sense minimal.

Obviously, while Cases 1 through 3 describe fairly general learning problems, Case 4 illustrates just one of the specific approaches to nonsupervised learning (For further information on these see, e.g., Cypkin (1970), Saridis, Nicolic and Fu (1969) and Vasil'ev (1967)).

When learning with a teacher, we have, in addition to the sequence $\{\omega_i ; i = 0,1,...\}$ of samples also an associated sequence of labels $\{\ell_i ; i = 0,1, ...\}$. However, this latter is absent in nonsupervised learning. Nevertheless, we assume in what follows in all problems also labels. This makes no matter if one thinks, in the case of nonsupervised learning, of a label sequence which equals, at any instant, the same constant.

Up to this point our learning model is still incomplete. Viz., there is as yet undefined (i) how to judge the appropriateness of the estimates, (ii) in what sense to average costs, etc. We have, as yet, given no description of how the samples are picked from Ω , and how are, in the case of ambiguous hypothesis testing, the features $\omega_i, i = 0,1,\ldots$ and the validity of the hypotheses H_0 and H_1 related. Obviously, these sort of questions may be answered in probabilistic terms. We are concerned with these in Sec. 1.2.

1.2 Probabilistic Notions

Let us adopt the hypothesis that $\zeta_i = (\omega_i, \ell_i)\,(i = 0, \pm 1, \ldots)$ are random variables defined in some probability space (Λ, F_Λ, P) Here Λ denotes the set of elementary events. Let $\Lambda = X_{i=-\infty}^{\infty}(\bar{\Omega}_i \times \tilde{\Omega}_i)$ where $\bar{\Omega}_i = \{\zeta_i\}$, and $\tilde{\Omega}_i$ offers, for any i , further possibilities to describe random perturbations and introduce randomization, as it will be actually done in the sequel. F_Λ denotes the set of considered events which is, by definition, a σ- algebra. $F_\Lambda = X_{i=-\infty}^{\infty}(F_{\bar{\Omega}_i} \times F_{\tilde{\Omega}})$. Here $F_{\bar{\Omega}_i}(i = 0,1,\ldots)$ denotes the σ-algebra, generated by sets of the form $(\zeta_i : \langle \zeta_i, \bar{e}_k \rangle \leq C$ $(\{\bar{e}_k\}_{k=0}^{\bar{N}}$ denoting a base in $\bar{\Omega}$, $F_{\tilde{\Omega}_i}$ is some σ-algebra of subsets of $\tilde{\Omega}_i$, $\bar{N} = \dim \Omega$, C any real, and $\langle .,. \rangle$ the Euclidean inner product in an $\bar{N}$ -variate $\bar{\Omega}$).

Obviously, by this description one may still consider any sort of dependence concerning the sequence $(\omega_i, \ell_i;\ i = 0, \pm 1, \ldots)$. However, in most cases of machine learning we do not entirely need this generality. One may, within a broad class of problems, fairly well assume that the label $\ell_n (n = 0, \pm 1, \ldots)$ depends on $(\omega_i, \ell_i;$ $i < n)$ only through ω_n. This is obviously the case if

$\ell_n = \Theta(\omega_n)$. However, for the sake of simplicity, we assume such a property also in Case 3, in the sequel, assuming that

$$(1.2.6) \qquad P(d_n = | \omega^n, d^{n-1}, \tilde{\omega}^{n-1}) = P(d_n = 1 | \omega_n) \text{ and } P(d_n = 1 | \omega_n = \omega) = \Pi(\omega),$$

independently of n . (We call, in what follows, $\Pi = \{\Pi(\omega), \omega \in \Omega\}$ a posteriori probability function.) Observe that here, and in what follows, the superscript refers to the past of the sequence considered, up to and including n . (I.e., $\omega^n = \{\omega_i, i \leq n\}$, etc.)

(1.2.6) is not a severe constraint, since actual systems usually have, with respect to past events, decaying memory. Therefore, by defining, e.g., $\omega_n = (\tilde{\rho}_i)^n_{i=n-n_0}$ (where $\tilde{\rho}_i$ is the quantity actually, observed at i), one may usually pick an n_0 for which (1.2.6) may be fairly well assumed. By this procedure, however, we obtain, even for a measurement sequence $\{\tilde{\rho}_i\}_{i=-\infty}^{\infty}$ which is totally independent, a dependent sample sequence $\{\omega_i\}_{i=-\infty}^{\infty}$.)

Dependence may play, because of inertia, an important role in many learning problems in which the samples $\omega_i (i = 0, 1, \ldots)$ characterize, at different instances, the performance (or behavior) of the very same real-life system. However, in many of the classical problems in pattern recognition $\{\omega_i\}_{i=0}^{\infty}$ may, obviously, be assumed totally independent. Specifically such problems are considered, in the sequel, by assuming that $\omega_i (i = 0, \pm 1, \ldots)$ are drawn from Ω , totally independently, according to some probability distribution Q .

In actual problems of machine learning different

kind of a priori knowledge may be assumed for P (Π and Q, eventually). However, typical are those pattern recognition problems for which only such general properties of $\{\omega_i\}_{i=-\infty}^{\infty}$ may be a priori assumed, as (i) independence, (ii) weak dependence (in the sense to be specified in the sequel) (iii) that certain moments exist or even have uniform bounds, etc. We are interested in what follows only in such sort of problems.

Having introduced such probabilistic hypotheses, machine learning may be formulated at any instant n as a problem of statistical inference from (ω^n, ℓ^n) on the very quantity which is considered as the target of learning.

Obviously in Cases 1, 2 and 3 we have a problem of statistical estimation. Viz., in Case 1 we have to propose at some (or perhaps each) instant n, a real valued estimate $X_n = \{X_n(\omega), \omega \in \Omega\}$ on the unknown function Θ. In Case 2 we may define at some n a real valued estimate X_n, but only $\mathrm{sign} X_n - \mathrm{sign}\,\Theta$ is of actual interest. Case 4 is specific in the respect, since here the target quantity Θ and its estimate X_n (at n) are not functions defined in Ω, but members of $\Omega \times \Omega$. In Case 4 the target quantity of learning is, specifically, $E\tilde{Y}_n$. (E stands for the expectation.)

In Case 3 it is, in many cases, reasonable to consider as the target quantity of learning such a decision function $\delta^* = \{\delta^*(\omega), \omega \in \Omega\}$ for which the probability $P(\delta_n \neq d_n)$ of misclassification is minimal, and devise, at some instant n, an estimate $\delta_n = \{\delta_n(\omega),\ \omega \in \Omega\}$ of this optimal decision function δ^*.

In Chapters 2–5 our subject is, how to devise such estimates, having meaningful properties.

1.3 Learning Optimal Decision Functions

In ambiguous hypothesis testing it is fairly natural to characterize ambiguity by the a posteriori probability function as defined by (1.2.6). Moreover, it is well known from classical theory how may one from Π derive an optimal decision function.

We reproduce in this secion the proof of this well known relation with a slight extension. Viz., we also admit for $\{\omega_i\}_{i=0}^{\infty}$ any sort of dependence, under the constraint that $\{\ell_i\}_{i=0}^{\infty}$ and $\{\omega_i\}_{i=0}^{\infty}$ are related according to (1.2.6).

Denote by δ some decision function, proposed for the classification of ω_n. Let $\delta(\Pi) = \{\delta(\Pi,\omega)\ \omega \in \Omega\}$ stand, specifically, for the so called standard decision function, defined by:

$$\delta(\Pi,\omega) = \begin{cases} 0\ , & \Pi(\omega) < 1/2\ , \\ 1\ , & \Pi(\omega) \geq 1/2\ . \end{cases} \tag{1.3.7}$$

Let

$$B_n = \left(\lambda : \delta(\omega_n) = 1\right)\ , \quad A = \left(\omega : \delta(\omega) = 1\right),$$

$$\tilde{B}_n = \left(\lambda : \delta(\Pi,\omega_n) = 1\right)\ , \quad \tilde{A} = \left(\omega : \delta(\Pi,\omega) = 1\right).$$

Assume, for the moment, that δ is used to classify ω_n. Observe that, specifically in this case ω_n is misclassified if $\delta(\omega_n) \neq d_n$. Thus the probability of misclassification obviously is

$$P_e(n,\delta) = \int_{B_n^c} \Pi(\omega_n) P(d\lambda) + \int_{B_n} (1 - \Pi(\omega_n)) P(d\lambda) \qquad (1.3.8)$$

(Here, and in what follows, superscript "c" stands for the complementary event: $B_n^c = \Lambda - B_n$.)

For a sample sequence $\{\omega_i\}_{i=-\infty}^{\infty}$ of arbitrary dependence the probability of misclassification depends, of course, on n. However, if ω_i, $i = 0, \pm 1, \ldots$, are drawn from Ω totally independently according to a probability distribution Q, we readily obtain from (1.3.8)

$$P_e(n,\delta) = \int_{A^c} \Pi(\omega) Q(d\omega) + \int_{A} (1 - \Pi(\omega)) Q(d\omega) \quad , \qquad (1.3.9)$$

independently of n (We may simply write $P_e(n,\delta) = P_e(\delta)$ for this case.)

<u>THEOREM 1.1</u> $P_e(n,\delta) \geq P_e(n, \delta(\Pi))$ for any $n = 0, 1, \ldots$; i.e. (1.3.7) is, for any n, an optimal decision function.

<u>Proof</u> Theorem 1.1 follows almost obviously from (1.3.7). We present the proof because it serves also as a use-

ful reference in the sequel. (1.3.8) may be rewritten as

$$P_e(n,\delta) = \int_\Lambda \left[\left(1-\chi_A(\omega_n)\right)\Pi(\omega_n) + \chi_A(\omega_n)\left(1-\Pi(\omega_n)\right)\right] P(d\lambda)$$

(1.3.10)
$$= P(d_n=1) + \int_\Lambda \chi_A(\omega_n)\left(1-2\,\Pi(\omega_n)\right) P(d\lambda)$$

(Here χ_A is the indicator of A and $P(d_n=1) = \int_\Lambda \Pi(\omega_n) P(d\lambda)$.)

From (1.3.10)

(1.3.11) $$P_e(n,\delta) - P_e(n,\delta(\Pi)) = \int_\Lambda \chi(\omega_n)\left(1-2\Pi(\omega_n)\right) P(d\lambda),$$

where $\chi = \chi_A - \chi_{\tilde{A}}$.

From (1.3.11) obviously follows

$$P_e(n,\delta) - P_e(n,\delta(\Pi)) \geq 0 .$$

(For this observe only that for $\chi(\omega_n) < 0$ we have $\chi_{\tilde{A}}(\omega_n) = 1$. However, $\chi_{\tilde{A}}(\omega_n) = 1$ iff $\Pi(\omega_n) > 1/2$ and $1-2\Pi(\omega_n) < 0$. Thus $\chi(\omega_n)\left(1-2\Pi(\omega_n)\right) \geq 0$. This completes the proof of Theorem 1.1.

<u>REMARK</u> In typical machine learning the a posteriori probability function Π is, of course, a priori unknown. Still knowing that $\delta(\Pi)$ is optimal is also of interest in this case, since it suggests us to estimate, at some (or all) $n = 0,1,\ldots,$ the a posteriori probability function Π by some real valued function $Z_n = \{Z_n(\omega), \omega \in \Omega\}$, and then propose, as δ_n, the decision function $\delta_n(Z_n)$, defined by

$$\delta_n(Z_\lambda,\omega) = \begin{cases} 0 \ , \text{ if } \ \bar{Z}_n \leq 1/2 \ , \\ 1 \ , \text{ if } \ \bar{Z}_n > 1/2 \ . \end{cases}$$

$\bar{Z}_n = Z_n$ if $Z_n \in [0,1]$, $\quad \bar{Z}_n = 1$ if $Z_n > 1$, and $\bar{Z}_n = 0$ if $Z_n < 0$,

This procedure is, of course, only then appropriate if we, specifically, have $Z_n \to \Pi$, as $n \to \infty$, in a sense which also guarantees the convergence of the probability of misclassification to $P_e(n, \delta(\Pi))$.

It turns out that one may devise, in this way, in a fairly straightforward manner algorithms for learning optimal decision functions.

According to this we consider, in what follows, specifically in Case 3, the estimation of such real valued functions as $\Pi = \{\Pi(\omega), \omega \in \Omega\}$ as the target of learning. (E.g., $\Theta = \Pi$) However $Z_n \to \Theta$ is only then meaningful if $P_e(n,\delta(Z_n)) \to P_e(n,\delta(\Theta))$ as $n \to \infty$.

1.4 Iterative Learning

In all illustrations, we gave in Sec. 1.2, the target quantity Θ of the considered learning problems was either some finite sequence of members of Ω, or some real valued function defined in Ω. (Observe, that in this latter case Θ may also be specified by a finite number of reals, provided Ω is, as in digital computation indeed, a finite set.) As a matter of fact in Cases 1 through 4 we had $\Theta \in \mathcal{H}$, $\mathcal{H}$ being some finite dimensional Hilbert space.

We have also seen in Sec. 1.2 situations in which one a priori knows that Θ is within some subset $\mathcal{A}$ of $\mathcal{H}$.

Accordingly, we complete our model by assuming that the target quantity Θ of any considered learning problem is within some $\mathcal{A} \subset \mathcal{H}$, $\mathcal{H}$ being some given Hilbert space, and $\mathcal{A}$ a given subset, but not necessarily a subspace, of $\mathcal{H}$.

For the sake of simplicity we assume, in the present context, that $\dim \mathcal{H} = N < \infty$. (However, most of what we say can also be proved for such an $\mathcal{H}$ which is separable. In theoretical studies it is actually reasonable to adopt also this further extention, in order to avoid, in models, undue descriptions.)

We denote by $\langle \xi, \eta \rangle$ the inner product, and by $\|\xi\| = \sqrt{\langle \xi, \xi \rangle}$ the norm, for any $\xi, \eta \in \mathcal{H}$.

As an estimate X_n, we admit any function measurable with respect to the data ω^n, ℓ^n, and the outcome $\tilde{\omega}^n$ of actual random experiments up to n. Let us denote this dependence by $X_n = X_n(\omega^n, \ell^n, \tilde{\omega}^n)$. (X_n is also assumed to take-values in $\mathcal{A}$.)

If we introduce no further constraints, such an estimate may equally well admit any sort of handling data (ω^n, ℓ^n). It is, in this respect, principally irrelevant wether (i) the samples ω^n are presented afteranother or, say, all at the same time (ii) we propose X_n as a final estimate (as in a fixed sample size problem) or it is just one of sequentially proposed estimates X_i, $i = 0,1,\ldots$ (from which the final one is picked by a stopping rule).

However, within the scope of the latter sequential procedures the size of the sample may grow excessive (i) because we arrive at the required quality of estimation too slowly, (and run out of storage capacity) or (ii) because we wish also to prolong the procedure (observe the variations of the system) unlimitedly.

It is a possible way out of this difficulty, to restrict the number of ω_i, ℓ_i from which X_n is actually computed, and store, instead of the excluded data, the most recent or a number of most recent estimates $X_i, i \leq n$. A simple choice is to have $X_n = X_n(\omega_n, \ell_n, X_{n-1})$. We term all these sort of procedures iterations.

In what follows we are specifically interested in iterative algorithms, which may, at any $n = 0,1,\ldots$, be still appropriately studied in terms of successive corrections with respect to the previous estimate.

The estimate of Θ at n is X_n . Let $\left\{X_n\right\}_{n=0}^{\infty}$ be an iteration sequence defined successively, starting at some arbitrary

$$X_0 \in \mathcal{A} \subset \mathcal{X} \left(E \left\| X_0 \right\|^2 < \infty\right).$$

by

$$X_{n+1} = \Phi\left(X_n + \alpha_n W_n(X^n)\right). \tag{1.4.12}$$

$W_n(X^n)=\{W_n(X^n,\omega),\omega\in\Omega\}$ denotes, for any $n=0,1,\ldots$, a random variable taking values in $\varkappa$. Let $\left(X^n=(\bar{X}_\nu,\nu\leq n)\right.$ where $\bar{X}_\nu = X_\nu$ for $0\leq\nu\leq n$ and $\bar{X}_\nu=0$ for $\left.\nu<0.\right)$ One may, by means of W_n, correct the iteration, at instant n, according to what was taught or evaluated up to n. We admit any W_n which is a measurable function of $\omega^{n+1}, \ell^{n+1}, \tilde{\omega}^{n+1}$, i.e. $W_n = W_n(\omega^{n+1}, \ell^{n+1}, \tilde{\omega}^{n+1})$. We call W_n regulator in the sequel.

$\{\alpha_n\}_{n=0}^{\infty}$ denotes an a priori specified sequence of positive reals, to be defined also in the sequel. (Provided $\alpha_n=\alpha_0$, for all n, it is just an irrelevant formalism, in (1.4.12), that $X_n+\alpha_n W_n$ is written. However, if we have such $\{\alpha_n\}_{n=0}^{\infty}$ which actually depends on n, this specifically means that, for any n, the regulator W_n is, a priori, weighted with respect to X_n. One may, in this way, obtain more freedom for devising the regulator and combating ambiguity and noise, at the price of loosing sensitivity.

$\Phi : \varkappa \to \mathcal{A}$ denotes a truncation. Provided $X_0\in\mathcal{A}$, one may keep, by means of Φ, $\{X_n\}_{n=0}^{\infty}$ within $\mathcal{A}$. (We show in Sec. 1.5 constraints, under which such truncation may be introduced with-

---

(*) We will occasionally also study the behavior of $W_n(Z^n)$ at some arbitrary sequence $Z=\{Z_n, n=0,\pm1,\ldots\}$ Because of this we introduce the notation $W(Z^n)$. However, specifically for $W_n(X^n)$ we will also use as a brief notation simply W_n.

out any difficulty.) We call $\mathcal{A}$ the domain of learning, in the sequel.

The main problem, in devising algorithms of the form (1.4.12), is to define appropriately the regulator W. In so doing one may consider regulators of various sort of dependence, concerning previously presented data. Many and widely used regulators are of the form: $W_n = W_n(\omega_{n+1}, \ell_{n+1}, \tilde{\omega}_{n+1})$. In such cases we say the regulator W (and the associated iteration rule) to be memoryless. Under more general conditions we speak of regulators and iteration rules with memory.

1.5 The Domain of Learning

If one has some a priori data, within which part of the space $\varkappa$ may Θ be found, it is reasonable to make use of this knowledge. A question of interest is, in this respect, under what constraint may the domain $\mathcal{A}$ of learning be confined to some proper subset of $\varkappa$, without loosing control of the convergence properties.

In this section we define a property by which such truncations may be introduced without any difficulty.

<u>DEFINITION</u> A truncation Φ is called with respect to some domain $\mathcal{A}$ of learning, uniformly norm reducing, if

$$\|\Phi(\xi) - \eta\| \leq \|\xi - \eta\|, \tag{1.5.1}$$

for any $\xi \in \mathcal{X}$ and $\eta \in \mathcal{A}$,

<u>LEMMA 1.1</u> Let $\mathcal{A}$ be a hypercube within $\mathcal{X}$, i.e.

$$(1.5.2) \quad \mathcal{A} = \left\{ \xi : \xi \in \mathcal{X};\ -\infty \leq a_i \leq \langle \xi, e_i \rangle \leq b_i \leq \infty,\ i = \overline{1,N} \right\}$$

and define the truncation as follows

$$\Phi(\xi) = \xi \quad \text{if} \quad \xi \in \mathcal{A} \quad ,$$

$$(1.5.3) \qquad \langle \Phi(\xi), e_i \rangle = \begin{cases} a_i & \text{if} \quad \langle \xi, e_i \rangle < a_i\,, \\ b_i & \text{if} \quad \langle \xi, e_i \rangle > b_i\,. \end{cases}$$

($N = \dim \mathcal{X}$, $\left\{ e_i \right\}_{i=1}^{N}$ is a base in $\mathcal{X}$, spanning the cube $\mathcal{A}$. a_i, $b_i (i = \overline{1,N})$ are reals, defining the vertices.) Then Φ is, with respect to $\mathcal{A}$, uniformly norm reducing.

<u>Proof</u> Obviously follows from (1.5.3). (Observe that, for any $i = \overline{1,N}$ and $\eta \in \mathcal{A}$, $a_i \leq \langle \eta, e_i \rangle \leq b_i$ and, therefore, $\left| \langle \Phi(\xi), e_i \rangle - \langle \eta, e_i \rangle \right| \leq \left| \langle \xi, e_i \rangle - \langle \eta, e_i \rangle \right|$.

<u>Remark</u> One may, adapting (1.5.3) (i) introduce, range delimitations with respect to a set of orthogonal coordinates, and (ii) dimensional constraints, (by setting $a_i = b_i$ for all $N_0 < i \leq N$, and (iii) search for Θ, at a time, only within a set of some subdivision of $\mathcal{X}$. Obviously, however, for successful learning, Θ has to be within $\mathcal{A}$.

CHAPTER 2

SOME ITERATION RULES WITHOUT MEMORY

2.1 Introduction

In this chapter we present some simple algorithms, governed by memoryless regulators. While these procedures are of actual interest also in their own right, they serve in the present context also to give some preliminary ideas in what form may converge constraints be formulated. One may get, in this way, more insight into how the notions and constraints, introduced in Chapter 3, are motivated.

In the present chapter we present theorems, omitting in most cases the proofs. However we supplement either all details or just the principles of also these proofs in Chapter 3.

2.2 Learning Governed by Costs

Assume we may, at any instant $n = 0,1,\dots,$ evaluate the performance of the learning procedure by some non-negative valued quantity $\tilde{Y}_n$, called cost.

This cost may have quite general sort of dependence, concerning what happened up to $n+1$. I.e., we may have $\tilde{Y}_n = \tilde{Y}_n(\omega^{n+1}, \ell^{n+1}, \tilde{\omega}^{n+1}, Z^n)$. We will be interested in looking also at $\tilde{Y}_n$ at some arbitrary $Z^n = \{Z_i, i \leq n\}(Z \in \mathcal{A})$. Therefore we introduce in this case for the cost function the partly abbreviated notation

$\tilde{Y}_n = \tilde{Y}_n(Z^n)$. If $Y_n(Z^n)$ really depends, given ω^{n+1}, ℓ^{n+1} and $\tilde{\omega}^{n+1}$ on Z^n and not just Z_n, we say the cost is with memory.

However there are many relevant problems, for which the cost $\tilde{Y}_n$ depends directly only on what happened at the very instant $n+1$. More distinctly: $\tilde{Y}_n = \tilde{Y}_n(\omega_{n+1}, \ell_{n+1}, \tilde{\omega}_{n+1} Z_n)$. We then say to have a memoryless cost, and such are the costs we consider in the present section. We will use for a memoryless cost function the notation $\tilde{Y}_n(Z_n)$.

Observe that by $Y_n(Z^n)$ and $Y(Z_n)$ we mean function defined in Ω. I.e.

$Y_n(Z^n) = \{\tilde{Y}_n(Z^n, \omega), \omega \in \Omega\}$ and $\tilde{Y}_n(Z_n) = \{\tilde{Y}_n(Z_n, \omega), \omega \in \Omega\}$, respectively.

<u>Example 1</u> $\tilde{Y}(X_n)$ may stand for the actual cost which is, e.g. financially experienced when observing ω_{n+1}.(This is the case of usual interest in many problems of business, industrial process control, etc.)

<u>Example 2</u> Assume that a function $\Theta = \{\Theta(\omega), \omega \in \Omega\}$ is taught, and how well X_n approximates Θ is, at instant $n+1$, evaluated by $\tilde{Y}_n(X_n, \omega_{n+1}) = (X_n(\omega_{n+1}) - \Theta(\omega_{n+1}))^2$.

<u>Example 3</u> Assume we wish to subdivide Ω into two classes, by means of learning without a teacher, in a way as pointed out in Case 4 (Sec. 1.1). Adopt for this purpose, at in-

stant $n+1$ the cost

$$\tilde{Y}_n = \chi_{A_n}(\omega_{n+1}) \left\| \omega_{n+1} - a_n \right\|^2 + \chi_{B_n}(\omega_{n+1}) \left\| \omega_{n+1} - b_n \right\|^2 .$$

(See (1.1.5)

Let

$$E\left(\tilde{Y}_n(X^n).|.\eta^n\right) = R(X_n) \qquad (2.2.1)$$

for any n (*). (E (.|.) stands for the conditional expectation).

We call, following usual statistical terminology, $R(\xi)$ the risk at ξ , and $R = \left\{R(\xi), \xi \in \chi\right\}$ the risk-function.

When such sort of costs $\left\{\tilde{Y}_n\right\}_{n=0}^{\infty}$ are available, it is quite natural to define the target of learning as follows: We seek for such a Θ , for which the risk is minimal, i.e.:

$$R(\xi) = \min ! \qquad (2.2.2)$$

if $\xi = \Theta$, with respect to all $\xi \in \mathcal{A}$.

In machine learning we usually introduce, apart from Example 1, $\tilde{Y}_n(\xi)$, and may therefore compute, at any n , either its gradient $\operatorname{grad}_\xi \tilde{Y}_n(.)$ (if this exists) or look for some other function which may take, in finding Θ , a similar role. (For the sake of simplicity, here we assume that $\operatorname{grad}_\xi \tilde{Y}_n(.)$ exists).

(*) Added in proof: This is a well known usual case if the sequence η is totally independent. For further extensions to weak dependence see Csibi (1973b,c).

Example If $\mathrm{grad}_\xi \tilde{Y}_n(\cdot) - (\xi(\omega) - \Theta(\omega))^2$ and $\xi(\omega) = \Sigma_{i=1}^N c_i e_i$ (For some base $\{e_i\}_{i=1}^N$) then

$$\langle \mathrm{grad}\, \tilde{Y}_n(., \omega_{n+1}), e_i \rangle \frac{\partial \tilde{Y}(\xi, \omega_{n+1})}{\partial c_i} = -2\Big(\Theta_n(\omega_{n+1}) - \sum_{i=1}^{\varkappa} c_i \varphi_i(\omega_{n+1})\Big)\varphi_i(\omega_{n+1})$$

(2.2.3)

for any $i = \overline{1,N}$.

In such cases it is well known and efficient way to use, for finding Θ, $Y_n(\xi, \omega_{n+1}) = \mathrm{grad}_\xi Y_n(., \omega_{n+1})$ instead of $\tilde{Y}_n$

Let $r(\xi) = \mathrm{grad}_\xi R$. Assume that we a priori know that (i) the risk function has, at least within $\mathcal{A}$, just a single minimum Θ , and (ii) we have here $r(\Theta) = 0$. (This is, obviously, the case in Example 2, even if $(.)^2$ is replaced by an arbitrary convex function $F(.)$).

The target of learning may be, in such cases, formulated in the following way: We seek for such a Θ, for which

$$r(\xi) = 0 \tag{2.2.4}$$

if $\xi = \Theta$, with respect to all $\xi \in \mathcal{A}$.

Observe that the mathematical problem, we arrived at, is to solve by means of observing $\{\tilde{Y}_n\}_{n=0}^{\infty}$, equations (2.2.2) resp. (2.2.4) which imply, however, moments we do not a priori know. These sort of problems are called, in Statistics, regression problems, and the equations to be solved regression equations. As a matter of fact, if a cost is specified, the task of iterative learning may be defined as solving some regression equation, recursively.

There is, of course, still considerable freedom in choosing iterations for this purpose. In this section we adopt (following essentially the approach of Newton's regula falsi but replacing R resp. r by its estimate $\tilde{Y}_n$ resp. Y_n) $W_n = W_n^{(0)} = -Y_n$, for any $n = 0,1,\ldots$.

We arrive in this way to an iteration process which is, specifically for $\Phi = I$, the well known Robbins Monro process. (I stands for the idem operator. Observe that solving regression problems, specifically by means of a Robbins Monro process, has originally been the starting point of what is called stochastic approximation. See e.g., Robbins and Monro (1951), Kiefer and Wolfowitz (1952), Dvoretzky (1956), Schmetterer (1960), Sakrison (1966), Cypkin (1968, 1970), Fu (1968) Ho and Agrewala (1968), and Wasan (1969).)

In iterative learning, governed by cost the very property may be for our help, in devising convergent algorithms, is to have a risk R which has, within $\mathcal{A}$, just a single minimum, no other extremum or inflexion point being approached within this domain. More distinctly, what we require in these cases is

<u>Condition 2.1</u> Let $Y = \{Y_n\}_{n=0}^{\infty}$ be such that there exists only a single $\Theta \in \mathcal{A}$, for which

$$\inf_{\varepsilon < \|\xi - \Theta\| < \varepsilon^{-1}} \langle \xi - \Theta, r(\xi) \rangle > 0 , \qquad (2.2.5)$$

for all $\varepsilon > 0$ and $\xi \in \mathcal{A}$.

Remark Constraint $\|\xi - \Theta\| < \varepsilon^{-1}$ admits the infimum to approach, as $\|\xi - \Theta\| \to \infty$, zero. For the convergence of iterations governed by a memoryless cost we have

THEOREM 2.1 If (i) $\Theta, X_0 \in \mathcal{A}$ (ii) $W = W^{(o)}$, $W^{(o)} = -Y_n$, where $Y_n = \mathrm{grad}_\xi \tilde{Y}_n$, ($\tilde{Y}_n (n = 0,1 \ldots)$ being memoryless costs) (iii) for Y_n, $n = 0,1,\ldots$, Condition 2.1 and $E(\|Y_n\|^2 | X^n) < C$ hold, (iv) the sequence $\{X_n\}_{n=0}^{\infty}$ of estimates is generated according to (1.4.12), with a truncation Φ which is uniformly norm reducing with respect to $\mathcal{A}$, and a sequence $\{\alpha_n\}_{n=0}^{\infty}$ for which $\alpha_n > 0$, for all n, and

$$(2.2.6) \qquad \sum_{n=0}^{\infty} \alpha_n = \infty, \qquad \sum_{n=0}^{\infty} \alpha_n^2 < \infty$$

holds, then $\lim_{n \to \infty} \|X_n - \Theta\| = 0$, almost surely.

Remarks (i) For the proof see Sec. 3.4. (ii) One has in many actual cases a priori knowledge concerning (2.2.5), having e.g. convex cost functions. However it also frequently happens, as it usually does in Case 4 in Sec. 1.1, that it is either difficult or, because of inherent properties of the considered problem, obviously impossible to guarantee (2.2.5). Nevertheless, if one knows that extrema and inflexion points are isolated, one may, by refining subdivisions in $\mathcal{X}$, arrive at an $\mathcal{A}$, around some extremum, for which (2.2.2) is met. One may devise, in this way, search procedures also for costs with a number of extrema and inflexion points. Obviously the use of such truncations, as Φ , has additional relevance in this respect.

(iii) While in classical regression problems one usually has an evidence for (2.2.5). However this is not the case in the aforementioned Example 3. As a matter of fact, while tasks of this sort are formulated as regression problems, the study of their iterative solutions is not based on the classical Theorem 2.1 devised for seeking for the root uniquely defined regression equations recursively, but on techniques described in Section 2.3. Typical learning problems related to Theorem 2.1 appear in the field of adaptive filters. (For a survey of such problems the Proakis and Miller 1969.)

2.3 Adopting Potential Function Type Kernels

Learning with a teacher as introduced in Chapter 1, is a sort of abstraction by which the labels given, together with each ω_i, $i = 0, \pm 1, \ldots$, by the teacher, are extended to the entire space Ω .

Let us think, for simplicity, only of an $\mathcal{X}$ which is a space of real functions defined in Ω . In this case it is a very natural way of making such an abstraction by the use of some a priori fixed interpolation function $K = \{K(\omega, \tilde{\omega}), (\omega, \tilde{\omega}) \in \Omega \times \Omega\}$

which takes real values.

More distinctly, let us, having received $(\omega_{n+1}, \ell_{n+1})$, correct the estimate X_n by placing this function at the very point ω_{n+1} in the sense that $W = W^{(1)}$, where

$$W_n^{(1)}(\omega) = \gamma_{n+1} K(\omega, \omega_{n+1}),$$

for all $\omega \in \Omega$ and $n = 0,1,\dots$. Here γ_{n+1} stands for a real valued quantity measurable with respect to $\omega_{n+1}, \ell_{n+1}, \tilde{\omega}_{n+1}$ and Z_n . I.e., $\gamma_{n+1} = \gamma_{n+1}(\omega_{n+1}, \ell_{n+1}, \tilde{\omega}_{n+1}, Z_n)$.

What happens in this case has some resemblance with the way how a point charge, placed to ω_{n+1} , generates a static field. This is while such kernels K are usually called potential functions, though hardly any further structural property supports this terminology. (Obviously, nothing relates K to the solution of a Laplace equation!).

A basic constraint on K , by which convergence of such algorithms may be guaranteed, is the following

Reproducing Property Assume K is such that

$$\langle \xi, K(., \omega) \rangle = \xi(\omega), \tag{2.3.1}$$

for any $\xi \in \varkappa$.

Remark 1 Assume Ω may be spanned by a finite set $\{\varphi_i\}_{i=1}^{N}$ of linearly independent functions defined in Ω . Accordingly, let us constraint the estimates $X_n (n = 0,1,\dots)$ to the finite linear manifold $\mathcal{M}$, spanned by $\{\varphi_i\}_{i=1}^{N}$. Define $\varkappa$ by $\mathcal{M}$, and the inner product as

$$\langle \xi , \eta \rangle = \sum_{i=1}^{N} c_i \tilde{c}_i , \tag{2.3.2}$$

for any $\xi = \sum_{i=1}^{N} c_i \varphi_i$ and $\eta = \sum_{i=1}^{N} \tilde{c}_i \varphi_i$.

It may be readily shown that, for $K(\omega, \omega_i) = \sum_{i=1}^{N} \varphi_i(\omega)\varphi_i(\omega_i)$ and $\varkappa$, the Reproducing Property holds. (This classical model of potential functions, introduced by Aizerman, Braverman and Rozonoer (1964, a, b, 1970) is of considerably actual interest because of the finite number of parameters it adopts).

Remark 2 A Hilbert space $\varkappa$ of functions, defined in Ω is called Reproducing Kernel Hilbert Space (RKHS) if there is a function $K = \{K(\omega,\tilde{\omega}), (\omega,\tilde{\omega}) \in \Omega \times \Omega\}$ for which $K(.,\omega) \in \varkappa (\forall \omega \in \Omega)$ and the Reproducing Property (2.3.1) holds. (See Aronszajn (1950)). It is well known that for any Hilbert space $\varkappa$ the Reproducing Property may hold for at most one function K , defined in $\Omega \times \Omega$. (We say, therefore, that such a K is the kernel of $\varkappa$.) On the other hand, there exists, for any positiv definite K , one and only one Hilbert space for which the Reproducing Property (2.3.1) holds. We call this Hilbert space the RKHS generated by K , and denote it by $H(K)$. From these relations follows that the theory of potential function type learning algorithms based on the Reproducing property, may be extended from finite dimensional to more general Hilbert spaces, provided a positive definite function K may be found for which $\Theta \in \mathcal{A} \in H(K)$. Basic facts, concerning such extentions and their consequences, are pointed out by Gulyas (1972).

Remark 3 For any pair of functions ξ and η in a RKHS $|\xi(\omega) - \eta(\omega)| \leq C_\omega \|\xi - \eta\|$ for any $\omega \in \Omega$. See Annex 2.

Let

(2.3.3) $$V_n^{(1)} = - E\left(\langle X_n - \Theta, W_n^{(1)} \rangle \,\middle|\, X^n \right).$$

Provided $\Theta \in \mathcal{A} \subset H(K)$, it follows from the Reproducing Property and the definition of $W_n^{(1)}$, that

(2.3.4) $$V_n^{(1)} = -E\left(\gamma_{n+1} \left(X_n(\omega_{n+1}) - \Theta(\omega_{n+1}) \right) \,\middle|\, X^n \right).$$

For potential function type learning algorithms γ is usually defined in a way that, for some $V_n^{(2)} \leq V_n^{(1)}$ the following two conditions hold:

Condition 1

$$V_n^{(2)} \geq 0 .$$

Condition 2

$$V_{n+1}^{(2)} \leq V_n^{(2)} + \alpha_n C_{10} + \zeta_n .$$

Here ζ_n is any sequence random variables, taking non-negative values, for which (i) ζ_n is, for any $n = 0, 1, \ldots$, measurable with respect to X^n and (ii) $\Sigma_{n=0}^{\infty} E\,\zeta_n < \infty$.

We are now in the position to present a basic convergence theorem for potential function type learning algorithms of this sort!

THEOREM 2.2 If (i) $\Theta \in \mathcal{A} \subset H(K)$ (ii) $W = W^{(1)}(W_n^{(1)} = \gamma_{n+1} K(\omega, \omega_{n+1}))$ (iii) $K(\omega, \tilde{\omega}) < C_1 < \infty$ (for all $(\omega, \tilde{\omega}) \in \Omega \times \Omega$ and some positive real C_1), (iv) $\{\gamma_n\}_{n=0}^{\infty}$ and K are such that Conditions 1 and 2 hold (v) the samples $\omega_n, n = 0, \pm 1, \dots$ are drawn from Ω completely independently, according to a probability distribution Q (vi) the sequence $\{X_n\}_{n=0}^{\infty}$ of estimates is generated according to the iteration rule (1.4.12) (vii) having a sequence $\{\alpha_n\}_{n=0}^{\infty}$ of coefficients, as given by (2.2.6), and adopting a uniformly norm reducing truncation Φ, <u>then</u> $\lim_{n\to} V_n^{(2)} = 0$, almost surely.

<u>Remark 1</u> For the proof see Sec. 3.5.

<u>Remark 2</u> It is, of course, still necessary to show that $V_n^{(1)}$ meaningfully describes the convergence of $\{X_n\}_{n=0}^{\infty}$ to Θ . We show this for some well known specific cases of $\{\gamma_n\}_{n=0}^{\infty}$ (Typical choices of the sequence $\{\gamma_n\}_{n=0}^{\infty}$, for which Theorem 2.2 holds, are presented in Cases A and B).

Remark 3 For the sake of simplicity, we are concerned within the present chapter only with potential function type algorithms without truncation. We are going to consider also potential function type algorithms with truncation in Chapter 4.

Case A Let us define, for learning a function Θ with a teacher

$$\gamma_n = \operatorname{sign}\left(\Theta(\omega_n) - X_n(\omega_n)\right) \tag{2.3.5}$$

From (2.3.4) and (2.3.5), taking the total independence of the presentation of $\{\omega_i\}_{i=0}^{\infty}$ into account,

$$V_n^{(1)} = E\left(\left|X_n(\omega_{n+1}) - \Theta(\omega_{n+1})\right| \Big| X^n\right) = \int_{\Omega} \left|X_n(\omega) - \Theta(\omega)\right| Q(d\omega). \tag{2.3.6}$$

which specifies a way of evaluating errors by the mean absolute value criterion.

For $V_n^{(2)} = V_n^{(1)}$ Condition 1, obviously, holds. Condition 2 follows from (C. 3) in Sec. 4.2, (1.4.12) and (iii) in Theorem 2.2. (Observe that (C.3) follows from (2.3.6), the Triangle Inequality and Appendix 2).

Case B Let us have, for ambiguous hypothesis testing, $\Theta = \Pi$ (See sec. 1.3.) Let

$$\gamma_n = \begin{cases} +1, & \text{if } d_n = 1, \mathcal{D}_n = 0, \\ 0, & \text{if } d_n = \mathcal{D}_n, \\ -1, & \text{if } d_n = 0, \mathcal{D}_n = 1, \end{cases} \tag{2.3.7}$$

where $\mathcal{D}_n$ is a random variable taking values 0 and 1 , according to

$$P\left(\mathcal{D}_{n+1} = 1 \mid \omega^{n+1}, d^{n+1}, \tilde{\omega}^{n+1}\right) = P\left(\mathcal{D}_{n+1} = 1 \mid \omega_{n+1}\right) = \bar{X}_n(\omega_{n+1}). \tag{2.3.8}$$

(Here $\bar{X}_n(\omega) = X_n(\omega)$ if $X_n(\omega) \in [0,1]$; $\bar{X}_n(\omega_{n+1}) = 1$, if $X_n(\omega) > 1$ and $\bar{X}_n(\omega) = 0$ if $X_n(\omega) < 0$.)

From (2.3.4) and (2.3.8)

$$V^{(1)} = -E\left(\gamma_{n+1}\left(X_n(\omega_{n+1}) - \Theta(\omega_{n+1})\right) \mid X^n\right) \geq \int_{\Omega} \left(\bar{X}_n(\omega) - \Theta(\omega)\right)^2 Q(d\omega) = V_n^{(2)} \tag{2.3.9}$$

Condition 1, this time again, obviously holds. Condition 2 follows from (C. 3), (1.4.12) and (iii) in Theorem (2.2). ((C.3) follows from (2.3.9) and (4.4.22)

$V_n^{(2)}$ may be considered, specifically, in the case of ambiguous hypothesis testing, meaningful only if it implies $P_e(n, \delta(X_n)) \to P_e(n, \delta(\Theta))$, as $n \to \infty$, provided any sample ω_{n+1} , $n = 0,1,\ldots$ is sorted according to the decision rule $\delta(X_n)$. (See

Sec. 1.3.) This property is actually guaranteed by the following

<u>LEMMA 2.1</u> If $\lim_{n\to\infty} E\left(\left(X_n(\omega_{n+1}) - \Theta(\omega_{n+1})\right)^2 \middle| \eta^n\right) = 0$, almost surely then $\lim_{n\to\infty}\left(P_e(n,\delta(X_n)) - P_e(n,\delta(\Theta))\right) = 0$ also almost surely.

<u>Remark 1.</u> Lemma 2.1 holds for any dependence of $\{\omega_n\}_{n=-\infty}^{\infty}$

<u>Remark 2.</u> For the proof see Appendix 1.

In Case B ambiguity and noise may also be involved. Even under such conditions the constraints of Theorem 2.2 guarantee convergence. (The main countermeasure against such fuzzyness is the use of a time-dependent sequence $\{\alpha_n\}_{n=0}^{\infty}$ of coefficients, according to (2.2.6). However the suppression of correction terms, in the course of learning, may make the algorithm inefficient if there is really no reason in so doing. It is therefore of interest to consider simple design constraints, under the condition of having $\alpha_n = \alpha_0$, for $n = 0,1,\ldots$ by which convergence may, specifically, for unambiguously defined learning problems be guaranteed, in some efficient way.

We consider in this respect two kind of learning models and according choices of $\{\gamma_n\}_{n=0}^{\infty}$.

Case C Let us consider, for learning a function Θ with a teacher,

$$\gamma_n = \gamma_n^{(0)} = \Theta(\omega_n) - X_n(\omega).$$

Case D Let us define, for unambiguous hypothesis testing,

$$\gamma_n = \gamma_n^{(1)} = \mathrm{sign}\,\Theta(\omega_n) - \mathrm{sign}\, X_n(\omega_n) \quad .$$

For Case C and D Györfi (1970) proved the following simple but interesting theorems

THEOREM 2.3 If (i) Conditions (i) - (iii) and (vi), given in Theorem 2.2, hold (ii) $\gamma_n = \gamma_n^{(0)}$ and (iii) $\alpha_n = \alpha_0 < 2/\sup_{(\omega,\tilde{\omega})\in\Omega\times\Omega} K(\omega,\tilde{\omega})$ for all n, (iv) $\Phi = I$, then $\lim_{n\to\infty}|X_n(\omega_{n+1}) - \Theta(\omega_{n+1})| = 0$, almost surely.

Proof Let: $\tilde{X}_n = X_n - \Theta$. From (1.4.12), and the Reproducing Property

$$\|\tilde{X}_{n+1}\|^2 \leq \|\tilde{X}_n\|^2 - \alpha_0\left(2-\alpha_0 K(\omega_{n+1},\omega_{n+1})\right)\left|\tilde{X}_n(\omega_{n+1})\right|^2 \qquad (2.3.10)$$

Observe that $\alpha_0\ 2-\alpha_0 K(\omega_{n+1},\omega_{n+1}) > C_{30} > 0$, and therefore $\|\tilde{X}_{n+1}\| \leq \|\tilde{X}_n\|$ and $\lim_{n\to\infty}\|\tilde{X}_n\|$, for all n, exists.

From (2.3.10)

$$E\|\tilde{X}_{n+1}\|^2 \leq E\|\tilde{X}_n\|^2 - C_{30}E|\tilde{X}_n(\omega_{n+1})|^2,$$

from which

$$C_{30}\sum_{n=0} E|\tilde{X}_n(\omega_{n+1})|^2 \leq E\|\tilde{X}_0\|^2 - E\|\tilde{X}_{n+1}\|^2 \leq E\|\tilde{X}_0\|^2 < \infty .$$

Thus $\lim_{n\to\infty}|\tilde{X}_n(\omega_{n+1})| = 0$, almost surely, which was to be proved.

THEOREM 2.4 <u>If</u> Conditions (i)-(iii), (vi) and (vii) given in Theorem 2.2 hold, (ii) $\gamma_n = \gamma_n^{(1)}$ (iii) we have, for any $\omega \in A\cup B$ and $\varepsilon > 0, |\Theta(\omega)| > \varepsilon$, and (iv) $\alpha_n > \alpha_0$ for all n , <u>then</u> $\lim_{n\to\infty}|\operatorname{sign} X_n(\omega_{n+1}) - \operatorname{sign}\Theta(\omega_{n+1})| = 0$ almost surely. (For A and B see Case 2 in Sec. 1.1)

<u>Remark</u> It is, specifically for Case D, previously known (see Aizerman, Braverman and Rozonoer (1964a) that

$$\|X_{n+1}\|^2 \leq \|X_n\|^2 - C_{40}|\operatorname{sign} X_n(\omega_{n+1}) - \operatorname{sign}\Theta(\omega_{n+1})|$$

From this point on the proof follows precisely what is used in proving Theorem 2.3.

2.4 Sequences of Kernels with Long Run Properties

An essential feature of the way we introduced, in Sec. 2.2, potential function type type algorithms, was the use of a step by step constraint for each γ_n and W_n, respectively.

Obviously, it is, just for getting convergence, not strictly necessary to adopt on W such step by step constraints.

We illustrate, in the present section, specifically for ambiguous hypothesis testing (See Case 3 in Section 1.1), another way of devising convergent learning algorithms.

Let us put the problem of ambiguous learning slightly in another way, as done previously. Consider, as in Case B, (Section 2.3) samples $\omega_n (n=0,1,\dots)$ drawn from Ω totally independently, according, at each n, to the same probability distribution Q. However, assume absolute continuity, in the sense that, for $j=0,1$ the probability density f_j, given the hypothesis H_j exists.

Take, instead of regarding Π as the target Θ of learning, at each ω, $\Theta(\omega) = p(\omega) = P(H_0)f_0(\omega) - P(H_1)f_1(\omega)$. It is well known that Π and p lead to the very same decision function if one accepts, for some ω, H_0 if $p(\omega) > 0$, and H_1, otherwise.

Looking at the learning problem in this way, offers new possibilities, however it makes also some specific constraints necessary.

Let us, also in this case, generate the learning algorithm by using a kernel, $k = \{k(\omega), \omega \in \Omega\}$ which is, however, now defined simply in Ω. Assume Ω to be, specifically an Euclidean M-space. Impose for this kernel the following

(2.4.1)
$$\begin{gathered} k(\omega) \geq 0, \\ k(\omega) \leq C_{40} < \infty, \text{ for all } \omega \in \Omega, \\ \int_\Omega k(\omega)\,d\omega = 1, \\ \int_\Omega \|\omega\|\, k(\omega)\,d\omega < \infty \quad , \end{gathered}$$

for all $\omega \in \Omega$.

Let $\varkappa = L_2$ with $\langle \xi, \eta \rangle = \int_\Omega \xi(\omega)\eta(\omega)\,d\omega$, for any $\xi, \eta \in L_2$. Accordingly:

(2.4.2)
$$\|X_n - \Theta\|^2 = I_n = \int_\Omega (X_n(\omega) - \Theta(\omega))^2\,d\omega .$$

Let

(2.4.3)
$$W_n^{(2)}(\omega) = (2d_{n+1} - 1)\,K_n(\omega, \omega_{n+1}) - X_n(\omega)$$

for all ω and $n = 0, 1, \ldots$. Here

(2.4.4)
$$K_n(\omega, \tilde{\omega}) = \frac{1}{h_n^M}\, k\left(\frac{\omega - \tilde{\omega}}{h_n}\right),$$

for all $(\omega, \tilde{\omega}) \in \Omega \times \Omega$,

$$h_{n+1} < h_n\,, \qquad \lim_{n\to\infty} h_n = 0, \qquad \lim_{n\to\infty} \frac{h_n^M}{\alpha_n}\,,$$

$$\sum_{n=0}^{\infty} h_n \alpha_n < \infty\,, \qquad \sum_{n=0}^{\infty} \frac{\alpha_n^2}{h_n^{-2M}} < \infty\,. \tag{2.4.5}$$

Assume specifically, that

$$\alpha_n = n^{-1}, \quad n \geq 1 \tag{2.4.6}$$

<u>THEOREM 2.5</u> <u>If</u> (i) $\Theta \in \mathcal{A} \subset L_2$ (ii) $W = W^{(2)}$ (iii) k, $\{h_n\}_{n=0}^{\infty}$ and $\{\alpha_n\}_{n=0}^{\infty}$ meet conditions (2.4.25) and (2.4.26) (iv) the samples $\omega_n (n = 0,1,\ldots)$ are drawn from Ω totally independently according, at each n , to the same absolute continuous probability distribution Q (v) assume the conditional probability density functions f_i , given H_i $(i = 0,1)$, is such that $|f_i(\omega) - f_i(\tilde{\omega})| < C_{40} \|\omega - \tilde{\omega}\|$, for all $(\omega, \tilde{\omega}) \in \Omega \times \Omega$, (vi) the sequence $\{X_n\}_{n=0}^{\infty}$ of estimates is generated according to the iteration rule (1.4.12), and (vii) we propose for the classification of any sample $\omega_{n+1} (n = 0,1,\ldots)$ the decision rule $\delta(X_n)$, <u>then</u> $\lim_{n\to\infty} I_n = 0$ <u>and</u> $\lim_{n\to\infty} \big(P_e(n+1, \delta(X_n)) - P_e(n+1, \delta(\Pi)\big) = 0$ almost surely.

<u>Remark</u> <u>(i)</u> We outline the proof in Sec. 3.6, referring to details in Wolverton and Wagner (1969), to whom Theorem 2.5 is due in its original form (having $\Phi = I$). <u>(ii)</u> The idea

imposing, for such interpolation functions, as K_n, convergence constraints on long run properties is due to Parzen (1962). (iii) Observe that optimal decision functions are learned, according to Case B, Sec. 2.3. (i), by adopting some additional randomization (ii) but the estimate is corrected, in this case, only if there is a disagreement between the classification d_{n+1}, given by the teacher, and $\mathcal{D}_{n+1}$, drawn by chance, according to an a posteriori probability defined by the most recent estimate X_n. In contrast to this, the algorithm, given by (2.4.3), involves no additional randomization, however the estimate is modified at each step in this case. – (iv) For both sort of learning algorithms it is a problem how to keep the complexity of the estimate under control. We are concerned, specifically with this topic in Chapter 5. (v) Theorem 2.5 is also instructive, as compared to Case B and Theorem 2.2, because of the way of posing the problem in terms of a moment $I_n (n = 0,1,...)$, with respect not to P but the apriori fixed Lebesgue measure of Ω, which is a direct consequence of considering p instead of Q.

CHAPTER 3

STABILITY OF STOCHASTIC PROCESSES

3.1 Introduction

In the present chapter we reproduce some recent Ljapunov type theorems, due to Braverman and Rozonoer (1969 a,b and 1970), on the stability of random processes.

We use, still within this chapter, these results to prove Theorems 2.1, 2.2 and 2.5. In addition, the same stuff also serves, in Chapter 4, as a reference to which more involved problems are reduced.

While these applications are of our actual interest, topics of the present chapter appear also instructive as an illustration of how Ljapunov type ideas on stability may directly be used in algorithmic design. The way machine learning problems are posed, offers particular possibilities to arrive, by using such sort of general approaches, at actual design conclusions.

3.2 Summary of Some Prerequisities

Let $\{\xi_n\}_{n=0}^{\infty}$ be a sequence of real valued random variables, defined on some probability space $\{\Lambda, F_\Lambda, P\}$. Recollect, as a further reference, on the convergence of $\{\xi_n\}_{n=0}^{\infty}$ the following two lemmas.

LEMMA 3.1 If $\lim_{n\to\infty}\xi_n=\mu$ in probability, then there exists such a subsequence $\{\xi_{n_i}\}_{i=0}^{\infty}$ for which $\lim_i \xi_{n_i}=\mu$ almost surely.

LEMMA 3.2 If $\lim_{n\to\infty}E|\xi_n|=0$ then $\lim_{n\to\infty}\xi_n=0$ in probability.

Let our interest be under what constraints does the sequence $\{\xi_n\}_{n=0}^{\infty}$ of random variables converge almost surely.

LEMMA 3.3 If $\sum_{n=0}^{\infty} E|\xi_n|<\infty$ then $\lim_{n\to\infty} \xi_n=0$, almost surely.

DEFINITION Let $\{\eta_n\}_{n=-\infty}^{\infty}$ be a sequence of random variables defined on $\{\Lambda,F_\Lambda,P\}$ and $\{\xi_n\}_{n=0}^{\infty}$ a sequence of real valued random variables measurable with respect to η_n . $\{\xi_n\}_{n=0}^{\infty}$ is a supermatringale (with respect to $\{\eta_n\}_{n=-\infty}^{\infty}$) if

(3.2.1) $$E(\xi_{n+1}|\eta^n) \le \xi_n ,$$

for any $n=0,1,\ldots$ (Recollect that $\eta^n=\{\eta_i, i\le n\}$. Replacing either $\le$ by $\ge$ or ξ_n by $-\xi_n$, we get a submartingale. For both cases the term semimartingale is used. For equality we say $\{\xi_n\}_{n=0}^{\infty}$ to be a martingale.)

Remark Let for some learning problem, with a target of learning Θ , $U_n=\|X_n-\Theta\|^2$ and $V_n=-E(\langle X_n-\Theta,W_n\rangle|\eta^n)$. From (1.4.12) and the uniformly norm reducing property of Φ , follows that

$$E\left(U_{n+1} \mid \eta^n\right) \leq U_n - 2\alpha_n V_n + \alpha_n^2 E\left(\|W_n\|^2 \mid \eta^n\right). \qquad (3.2.2)$$

Obviously if $\{\alpha_n\}_{n=0}^{\infty}$ is as given by (2.2.6), $V_n \geq 0$ and $E\left(\|W_n\|^2 \mid \eta^n\right) \leq C_0$ then $\{U_n\}_{n=0}^{\infty}$ approaches, as $n \to \infty$, to a supermartingale. As a matter of fact (3.2.2) makes the theory of such processes of interest within the problems of iterative learning, we consider in the present context.

<u>LEMMA 3.4</u> If ξ_n is a supermaratingale and $E\xi_0 \leq \infty$, then $E\xi_0 > E\xi_1 > \ldots > E\xi_n \ldots$

<u>Proof</u> Obviously follows from (3.2.1)

<u>THEOREM 3.1</u> If $\{\xi_n\}_{n=0}^{\infty}$ is a semimartingale, for which $E|\xi_n| < C_2 < \infty$ for all $n = 0,1, \ldots$, then there exists a random variable μ such that $\lim_{n \to \infty} \xi_n = \mu$, almost surely.

<u>Remark</u> For the proof see Theorem 4.1a, in Doob (1953).

<u>Remark</u> For further reference on prerequisites in the theory of martingales see such standard texts and monographs as Doob (1953), Loéve (1955), Neveau (1964), Feller (1966) and Širjaev (1969), etc.

3.3 Four Ljapunov Type Stability Theorems

We reproduce in this section some results due to Braverman and Rozonoer (1969a; 1970), concerning the convergence of certain sequences $\{G_n\}_{n=0}^{\infty}$, $\{U_n\}_{n=0}^{\infty}$ and $\{V_n\}_{n=0}^{\infty}$ of random variables, any member G_n, U_n, V_n of which is measurable with respect to η^n. $\{\eta_n\}_{n=0}^{\infty}$ is the sequence, the stability of which is considered. Obviously, one may also term in this case G_n, U_n, V_n functionals of η^n.

Let us use, in the present context, the term stability simply to denote that $\{\eta_n\}_{n=0}^{\infty}$ converges, as $n \to \infty$ in some meaningful sense, to some random variable, eventually to a constant, say, zero. We speak about Ljapunov type stability criteria if such a stability of $\{\eta_n\}_{n=0}^{\infty}$ is guaranteed in terms of constraints on one or, jointly a pair of functionals of η^n.

<u>THEOREM 3.2</u> Let (i) $\{\eta_n\}_{n=-\infty}^{\infty}$ be a sequence of random variables, taking values in an arbitrary space (ii) $\{G_n\}_{n=0}^{\infty}$ be a sequence of non-negative valued random variables, measurable at any n, with respect to $\eta^n = \{\eta_i, i \leq n\}$ (ii) $EG_0 < \infty$ and $E(G_{n+1} | \eta^n) \leq G_n + \varsigma_n$, for all $n = 0,1,\ldots$ and a sequence ς_n of non-negative valued random variables,(*) measurable with respect to η^n,

(*) Added in Proof: For extending studies of Chapters 3 and 4 to quasi-supermartingales with real valued but centered perturbations ς and related learning processes, see Csibi (1973b,c).

and such that $\sum_{n=0}^{\infty} E\xi_n<\infty$. Then there exists a random variable μ for which $\lim_{n\to\infty} G_n=\mu$, almost surely.

Proof Let

$$\tilde{G}_n = G_n + \sum_{i=n}^{\infty} \xi_i \tag{3.3.1}$$

Then

$$E\tilde{G}_0 \le E\left(G_0+\sum_{i=0}^{\infty}\xi_i\right) = EG_0+\sum_{i=0}^{\infty}E\xi_i < \infty \ . \tag{3.3.2}$$

From (3.3.1) and (3.3.2)

$$E\left(\tilde{G}_{n+1}\middle|\eta^n\right) = E\left(G_{n+1}\middle|\eta^n\right) + \sum_{i=n+1}^{\infty}\xi_i \le G_n+\sum_{i=n}\xi_i = \tilde{G}_n \tag{3.3.3}$$

Thus $\left\{\tilde{G}_n\right\}_{n=0}^{\infty}$ is a non-negativ valued supermartingale. From (3.3.2) we have $E\tilde{G}_0<\infty$ and therefore $E\tilde{G}_n< C_{21}$, for all $n=0,1,\ldots$ (see (3.3.3). It follows, therefore, from Theorem 3.1 that there exists a random variable μ for which $\lim_{n\to\infty} G_n=0$, almost surely. From this, however, it follows that $\lim_{n\to\infty} G_n=0$, almost surely, which completes the proof of Theorem 3.2.

THEOREM 3.3 (i) Let $\left\{\eta_n\right\}_{n=-\infty}^{\infty}$ be a sequence of random variables, taking values in some arbitrary measurable space, and (ii) $\left\{U_n\right\}_{n=0}^{\infty}$ and $\left\{V_n\right\}_{n=0}^{\infty}$ a pair of sequences of random variables, measurable, at any n , with respect to η^n . (iii) Let $U_n\ge 0$ $V_n\ge 0$ for all η .(iv) Assume that $EU_0<0$, (v) Let $\left\{U_n\right\}_{n=0}^{\infty}$ and $\left\{V_n\right\}_{n=0}^{\infty}$ be such that, for any n ;

$$E\left(U_{n+1}\middle|\eta^n\right) \le (1+\nu_n)\,U_n-\beta_n V_n+\xi_n . \tag{3.3.4}$$

Here $\{\nu_n\}_{n=0}^{\infty}$ is an a priori fixed real-valued, and $\{\beta_n\}_{n=0}^{\infty}$ a non-negative valued sequence of reals, and $\{\xi_n\}_{n=0}^{\infty}$ a non-negative valued sequence of random variables measurable, at any n, with respect to η^n, such that

$$(3.3.5) \qquad \lim_{n\to\infty}\beta_n=0, \ \sum_{n=0}^{\infty}\beta_n=0, \ \sum_{n=0}^{\infty}|\nu_n|<\infty, \ \sum_{n=0}^{\infty}E\xi_n<\infty.$$

(vi) In addition assume between $\{U_n\}_{n=0}^{\infty}$ and $\{V_n\}_{n=0}^{\infty}$ the following relation: For any sequence $\{n_i\}_{i=0}^{\infty}$ of subscripts, for which $\lim_{i\to\infty} V_{n_i}=0$ almost surely, we have $\lim_{i\to\infty} U_{n_i}=0$ in probability. If these conditions hold, then $\lim_{i\to\infty} U_n=0$, almost surely.

<u>Proof</u> The proof consists of two parts. First we show, by means of Theorem 3.2, that $\{U_n\}_{n=0}^{\infty}$ approaches, as $n\to\infty$, to some random variable μ, almost surely. Next we prove that $\mu=0$ also almost surely. (In reproducing the proof we follow an improvement due to Győrfi (1970).)

<u>Proving convergence to some random variable</u> Let

$$(3.3.6) \qquad G_n=U_n\prod_{i=n}^{\infty}\left(1+|\nu_i|\right)$$

and

$$(3.3.7) \qquad C=\prod_{i=0}^{\infty}\left(1+|\nu_i|\right)>1$$

Obviously

$$\frac{1}{C} G_n \leq U_n \leq G_n . \tag{3.3.8}$$

From (3.3.6), (3.3.4) and (3.3.7) follows that

$$\begin{aligned} E\left(G_{n+1} \mid \eta^n\right) &= E\left(U_n \mid \eta\right) \prod_{i=n+1}^{\infty} \left(1 + |\nu_i|\right) \leq \\ &\leq \left((1+\nu_n)U_n - \beta_n V_n + \varsigma_n\right) \prod_{i=n+1}^{\infty} \left(1 + |\nu_i|\right) \leq \\ &\leq \prod_{i=n}^{\infty} \left(1 + |\nu_i|\right) U_n - \beta_n V_n + C \varsigma_n \\ &= G_n - \beta_n V_n + C \varsigma_n . \end{aligned} \tag{3.3.9}$$

Thus:

$$E\left(G_{n+1} \mid \eta^n\right) \leq G_n + C \varsigma_n . \tag{3.3.10}$$

From (3.3.6):

$$E G_0 = E\left(\prod_{i=0}^{\infty} \left(1 + |\nu_i|\right) U_0\right) = E U_0 \prod_{i=0}^{\infty} \left(1 + |\nu_i|\right) < \infty . \tag{3.3.11}$$

Since, for $\{G_n\}_{n=0}^{\infty}$ all conditions of Theorem 3.3 are met, there exists some random variable μ for which $\lim_{n\to\infty} G_n = \mu$, almost surely. Observe that, by this (3.3.6) and (3.3.5)

$$\lim_{n\to\infty} U_n = \lim_{n\to\infty} \frac{G_n}{\prod_{i=n}^{\infty}\left(1 + |\nu_i|\right)} = \frac{\lim_{n\to\infty} G_n}{\lim_{n\to\infty} \prod_{i=n}^{\infty}\left(1 + |\nu_i|\right)} = \lim_{n\to\infty} G_n = \mu , \tag{3.3.12}$$

almost surely.

<u>Completing the proof</u> We have still to show that $\mu=0$. From

$$EG_{n+1} = E\left(E\left(G_{n+1}\middle|\eta^n\right)\right) < EG_n - \beta_n EV_n + CE\xi_n \tag{3.3.13}$$

and, therefore

$$EG_{n+1} \leq EG_0 - \sum_{i=0}^{n} \beta_i EV_i + C\sum_{i=0}^{n} E\xi_n \ . \tag{3.3.14}$$

Observe that $G_{n+1} \geq 0$. From this and (3.3.15)

$$0 \leq EG_0 - \sum_{i=0}^{n} \beta_i EG_i + C\sum_{i=0}^{n} E\xi_i , \tag{3.3.15}$$

for all n . Since $\sum_{n=0}^{\infty} E\xi_n < \infty$, (3.3.16) may only hold if

$$\sum_{n=0}^{\infty} \beta_n EV_n < \infty \tag{3.3.16}$$

Observe that $\sum_{n=0}^{\infty} \beta_n < \infty$ and $EV_n \geq 0$. From this and (3.3.16) follows, that there is some subsequence $\{n_i\}_{i=0}^{\infty}$ of the subscripts $n = 0,1,\ldots$,for which

$$\lim_{i\to\infty} EV_{n_i} = 0 \ , \tag{3.3.17}$$

From (3.3.17) and Lemma 3.2 follows $\lim_{i\to\infty} V_{n_i} = 0$ in probability. Thus, by Lemma 3.1, there exists a subsequence $\{n_{i_k}\}_{k=0}^{\infty}$ for which $\lim_{k\to\infty} V_{n_{i_k}} = 0$ almost surely. From this and Condition (vi) of Theorem 3.3 follows $\lim_{k\to\infty} U_{n_{i_k}} = 0$, in probability. However, we have already shown in the first part of the proof (see 3.3.12) that there exists a random variable μ for which $\lim_{n\to\infty} U_n = \mu$, almost surely.

From these last two relations, follows $\mu = 0$ i.e., $\lim_{n\to\infty} U_n = 0$ almost surely which has to be proved.

<u>THEOREM 3.4</u> (i) Let $\{\eta_n\}_{n=-\infty}^{\infty}$ be a sequence of random variables, taking values in some arbitrary measurable space, (ii) $\{U_n\}_{n=0}^{\infty}$ and $\{V_n\}_{n=0}^{\infty}$ a pair of sequences of random variables measurable, at any n, with respect to η^n. (iii) Let $U_n \geq 0$ and $V_n \geq 0$, for all n. (iv) Assume that $EU_0 < \infty$. (v) Let $\{U_n\}_{n=0}^{\infty}$ and $\{V_n\}_{n=0}^{\infty}$ be related, at any n, according to (3.3.4). (vi) In addition, assume that, for V_n, the following inequality holds, at any n, almost surely:

$$V_{n+1} \leq (1 + \bar{A}\beta_n)V_n + \bar{B}\beta_n + \varsigma_n, \tag{3.3.18}$$

Here $\bar{A}$ and $\bar{B}$ are a priori fixed non-negative reals, $\{\varsigma_n\}_{n=0}^{\infty}$ denotes a sequence of non-negative valued random variables, for which $\sum_{n=0}^{\infty} E\varsigma_n < \infty$. If these conditions hold, then $\lim_{n\to\infty} V_n = 0$, almost surely.

<u>Remark</u> We prove Theorem 3.4, referring to the first part of the proof of Theorem 3.3 and Lemma 3.6.

<u>LEMMA 3.6</u> <u>If</u> for some sequence $\{\rho_n\}_{n=0}^{\infty}$ of non-negative reals

(3.3.19) $$\sum_{n=0}^{\infty} \beta_n \rho_n < \infty$$

and

(3.3.20) $$\rho_{n+1} \leq (1+\beta_n)\rho_n + \beta_n + \tilde{\alpha}_n,$$

holds (where $\{\beta_n\}_{n=0}^{\infty}$ and $\{\tilde{\alpha}_n\}_{n=0}^{\infty}$ stand for sequences of non-negative reals, such that $\lim_{n\to\infty} \beta_n = 0$, $\sum_{n=0}^{\infty} \beta_n = \infty$ and $\sum_{n=0}^{\infty} \tilde{\alpha}_n < \infty$) then $\lim_{n\to\infty} \rho_n = 0$.

Proof Because of (3.3.19) we may, instead of (3.3.20), equally well consider

(3.3.21) $$\rho_{n+1} \leq \rho_n + \beta_n + \tilde{\tilde{\alpha}}_n$$

where $\tilde{\tilde{\alpha}}_n = \tilde{\alpha}_n + \beta_n \rho_n$. (Observe that $\sum_{n=0}^{\infty} \tilde{\tilde{\alpha}}_n < \infty$)

Let us, for any $\varepsilon > 0$, take the following set of all non-negative integers

(3.3.22) $$\Gamma(\varepsilon) = \left\{ n : \rho_n > \frac{\varepsilon}{2} \right\} = \left\{ n_i \right\}_{i=0}^{\tilde{N}}$$

$(n_i \leq n_{i+1})$. Provided $\tilde{N} < \infty$, Lemma 3.6 obviously holds. We may, therefore, confine ourselves to $\tilde{N} = \infty$, in the sequel.

Observe that, for $\tilde{N} = \infty$ and any $\varepsilon > 0$,

(3.3.23) $$\infty > \sum_{i=0}^{\infty} \beta_{n_i} \rho_{n_i} > \frac{\varepsilon}{2} \sum_{i=0}^{\infty} \beta_{n_i} .$$

Thus $\sum_{i=0}^{\infty} \beta_{n_i} < \infty$ I.e. there exists, for any $\varepsilon > 0$, such a non-negative integer ℓ for which

$$\sum_{i>\ell} \beta_{n_i} < \frac{\varepsilon}{4} . \tag{3.3.24}$$

Since $\lim_{n\to\infty} \beta_n = 0$ and $\sum_{n=0}^{\infty} \tilde{\tilde{\alpha}}_n < \infty$ (therefore $\sum_i^{\infty} \alpha_{n_i} < \infty$) there exists, for any $\varepsilon > 0$ a non-negative integer for which

$$\beta_n < \frac{\varepsilon}{8}, \tag{3.3.25}$$

for all $n \geq m$, and

$$\sum_{n>m} \tilde{\tilde{\alpha}}_n < \frac{\varepsilon}{8} . \tag{3.3.26}$$

Observe that $\sum_{n=0}^{\infty} \beta_n = \infty$, however $\sum_{i=0}^{\infty} \beta_{n_i} < \infty$, for $n_i \in \Gamma(\varepsilon)$ (which is assumed to be infinite). Therefore $|\Gamma(\varepsilon)^c| = \infty$ ($|S|$ denotes the number of members of a finite or enumerable set S. S^c is the complementary set of S. (Thus there exists a non-negative integer $\tilde{\tilde{N}} \in \Gamma(\varepsilon)^c$, for which $\tilde{\tilde{N}} > \max(\ell, m)$).

Let us now show that, for any $n_i \in \Gamma(\varepsilon)$, for which $n_i > \tilde{\tilde{N}}$, we have $\rho_{n_i} < \varepsilon$. (Observe that by so doing Lemma 3.6 is actually proved!)

Let for any such n_i

$$\tilde{n}_i = \max(n : n \in \Gamma(\varepsilon)^c,\ n > \tilde{N},\ n < n_i). \tag{3.3.27}$$

By (3.3.21)

(3.3.28) $$\rho_{n_i} \leq \rho_{\tilde{n}_i} + \beta_{\tilde{n}_i} + \sum_{i=\tilde{n}_i+1}^{n_i-1} \beta_i + \sum_{i=\tilde{n}_i}^{n_i-1} \tilde{\tilde{\alpha}}_i \ .$$

From (3.3.22), (3.3.24), (3.3.25) and (3.3.26)follows that $\rho_{n_i} < \varepsilon$ and, therefore, $\rho_n < \varepsilon$ for all $n > \tilde{N}$. This completes proof of Lemma 3.6.

Proof of Theorem 3.4 Up to (3.3.16) the proof of Theorem 3.3 obviously holds. As a result of this we have also for the present case

(3.3.29) $$\sum_{n=0}^{\infty} \beta_n \, EV_n < \infty \ .$$

From Markov's inequality it follows that, for any $\Delta > 0$ there exists some $C_\Delta > 0$ for which:

(3.3.30) $$P\left(\lambda : \sum_{n=0}^{\infty} \beta_n V_n(\lambda) < C_\Delta\right) > 1 - \Delta \ .$$

$(\lambda \in \Lambda.)$ Let, for any such λ and $n = 0,1,\ldots,$

(3.3.31) $$\rho_n = V_n(\lambda)$$

Obviously, for such $\{\rho_n\}_{n=0}^{\infty}$, Lemma 3.6 holds, from which $\lim_{n\to\infty} V_n(\lambda) = 0$ follows. Since this conclusion holds for any $\Delta > 0$, we have $\lim_{n\to\infty} V_n = 0$, almost surely. This completes the proof of Theorem 3.4.

<u>THEOREM 3.5</u> Let $\{\eta_n\}_{n=0}^{\infty}$ be a sequence of random variables, taking values in some arbitrary measurable space, (ii) $\{U_n\}_{n=0}^{\infty}$ and $\{V_n\}_{n=0}^{\infty}$ a pair of sequences of random variables measurable, at any n, with respect to η^n. (iii) Let $U \geq 0$ and $V \geq 0$ for all n. (iv) Assume that $EU_0 < \infty$ (v) Let $\{U_n\}_{n=0}^{\infty}$ and $\{V_n\}_{n=0}^{\infty}$ be related, at any n, according to (3.3.4), <u>having specifically $\beta_n = C$</u>. If these conditions hold, then $\lim_{n\to\infty} V_n = 0$, almost surely.

<u>Proof</u> Up to (3.3.16) the proof of Theorem 3.3 holds precisely also for this case. From (3.3.16) and $\beta_n = C$ for all n, follows

$$\sum_{n=0}^{\infty} EV_n < \infty .$$

From this and Lemma 3.3 $\lim_{n\to\infty} V_n = 0$ is obtained, almost surely, which was to be proved.

3.4 Application in Iterative Learning Regulated by Costs

Let

$$V_n^{(0)} = -E\left(\langle X_n - \Theta, W_n^{(0)}\rangle \middle| \eta^n\right) \tag{3.4.1}$$

and adopt the following substitutions

$$U_n = \left\| X_n - \Theta \right\|^2 \tag{3.4.2}$$

$$V_n = V_n^{(0)} \tag{3.4.3}$$

Using these notations, the basic Condition 3.1 on costs reads, as follows:

(3.4.4)
$$\inf_{\varepsilon \leq \|X_n - \Theta\| < \varepsilon^{-1}} V_n > 0,$$

for all n and any $\varepsilon > 0$.

<u>LEMMA 3.7</u> Condition (vi) in Theorem 3.3 holds if

(3.4.5)
$$\inf_{\varepsilon < \|\xi - \Theta\| < \varepsilon^{-1}} \langle \xi - \Theta, r(\xi) \rangle > 0,$$

for any $\xi \in \mathcal{A}$ and $\varepsilon > 0$.

<u>Proof</u> Here, and in what follows, we use the following abbreviation:

(3.4.6)
$$\tilde{X}_n = X_n - \Theta .$$

We prove more then Lemma 3.7, viz. show that $\lim_{n \to \infty} \|\tilde{X}_{n_i}(\lambda)\| = 0$ for any sequence $\{n_i\}_{i=0}^{\infty}$ and sample function $X(\lambda)$, for which $\lim_{n \to \infty} V_{n_i}(\lambda) = 0$.

Assume that, to the contrary of Lemma 3.7, there is some sequence $\{n_i\}_{i=0}^{\infty}$ and $\lambda \in \Lambda$ for which $\lim_{i \to \infty} V_{n_i}(\lambda) = 0$, however $\lim_{i \to \infty} \|\tilde{X}_{n_i}(\lambda)\| \neq 0$. In this case there exists a real $\tilde{\varepsilon} > 0$ and a sequence $\{n_{i_k}\}_{k=0}^{\infty} \in \{n_i\}_{i=0}^{\infty}$ for which $\|\tilde{X}_{n_{i_k}}(\lambda)\| > \tilde{\varepsilon}$ for all $n_{i_k} \geq C_{100}$.

Then, however, it follows from the definition of r and $W^{(0)}$ and (3.4.5) that $|V_{n_{i_k}}(\lambda)| \geq \tilde{q}(\tilde{\varepsilon})$ ($\tilde{q}(\tilde{\varepsilon})$ denoting for, any given $\varepsilon > 0$, some fixed positive real) which contradicts the initial assumption. By this Lemma 3.7 is proved.

Proof of Theorem 2.1 Let us adopt the following substitutions: $U_n = \|\tilde{X}_n\|^2$, $V_n = V_n^{(0)}$, $\beta_n = \alpha_n$, $\varsigma_n = \alpha_n^2 E(\|W_n^{(0)}\|^2 | \eta^n)$ and $\nu_n = 0$. Then (3.3.4) follows from (1.4.12), $W_n = W_n^{(0)}$ and (3.2.2), and Condition (vi) in Theorem 3.3 from (2.2.5) and Lemma 3.7. Observe that $U_n = \|\tilde{X}_n\|^2$ and $V_n = V_n^{(0)}$ meet all conditions in Theorem 3.3, from which $\lim_{n\to\infty} \|\tilde{X}_n\| = 0$, almost surely, follows. This completes the proof of Theorem 2.1.

3.5 Application in Learning Governed by Potential Functions

We owe only with the

Proof of Theorem 2.2 Recollect the Conditions enlisted in Theorem 2.2. Adopt the following substitutions: $U_n = \|\tilde{X}_n\|^2$, $V_n = V_n^{(1)}$, $\beta_n = \alpha_n$, $\varsigma_n = \alpha_n^2 E(\|W_n^{(1)}\|^2 | \eta^n)$ and $\nu_n = 0$. Observe that U_n and V_n meet all conditions in Theorem 3.4, from which $\lim_{n\to\infty} V_n^{(1)} = 0$, almost surely follows; by which Theorem 2.2 is proved.

3.6 Application in the Theory of Kernel Sequences with Long Run Properties

We only give an outline of the proof of Theorem 2.5, in order to show at which point do the ideas involved join the approach presented previously. For omitted details we refer to Wolverton and Wagner (1969).

Concerning the proof, it makes no essential difference that we admit, within the present context, also a uniformly norm reducing truncation Φ , in contrast to $\Phi = I$, considered in the original text.

(3.6.1) $$U_n = \|X_n\|^2 = \int_\Omega \tilde{X}_n(\omega)^2\, d\omega,$$

From (1.4.12) and the uniformly norm reducing property of Φ with respect to $\mathcal{A}$, follows:

(3.6.2) $$U_{n+1} = \int_\Omega \tilde{X}_{n+1}(\omega)^2 d\omega \leq \int_\Omega \tilde{X}_n^2(\omega)\, d\omega + 2\alpha_n \int_\Omega \tilde{X}_n(\omega)\, W_n^{(2)}(\omega)\, d\omega + \alpha_n^2 \int_\Omega \left(W_n^{(2)}(\omega)\right)^2 d\omega,$$

Observe that

(3.6.3) $$W_n^{(2)}(\omega) = L_n(\omega) - X_n(\omega)$$

where

(3.6.4) $$L_n(\omega) = \left(2d_{n+1}\right) K_n(\omega, \omega_{n+1}).$$

Adopt the following rearrangements:

(3.6.5) $$\begin{aligned} \tilde{X}_n(\omega)\, W_n^{(2)}(\omega) &= \tilde{X}_n(\omega)\, L_n(\omega) - \tilde{X}(\omega)\, X_n(\omega) \\ &= \tilde{X}_n(\omega)\left(L_n(\omega) - \Theta(\omega)\right) + \tilde{X}(\omega)\Theta(\omega) - \tilde{X}_n(\omega)\, X_n(\omega) \\ &= \tilde{X}_n(\omega)\left(L_n(\omega) - \Theta(\omega)\right) - \tilde{X}_n^2(\omega). \end{aligned}$$

From (3.6.2), (3.6.1) and (3.6.5)

$$U_{n+1} < U_n(1-2\alpha_n)+S_n$$

Here:

$$S_n = 2\alpha_n \int_\Omega \tilde{X}_n(\omega)\big((2d_{n+1}-1)\,K_n(\omega,\,\omega_{n+1})-\Theta(\omega)\big)\,d\omega+$$
$$+\,\alpha_n^2 \int_\Omega \big((2d_{n+1}-1)K_n(\omega\,,\omega_{n+1})-X_n(\omega)\big)^2 d\omega$$

The basic consequence of choosing $W_n = W_n^{(2)}$ is that $\sum_{n=0}^{\infty} ES_n < \infty$. However, we omit showing this, and just refer to Lemma 5 in Wagner and Wolverton (1969).

Obviously, adopting $U_n = V_n$, all conditions of Theorem 3.3 are met, from which $\lim_{n\to\infty} U_n = 0$, almost surely; from which the first assertion of Theorem 2.5 follows.

The proof of the second assertion (in Theorem 2.5) needs only a re-interpretation of the ideas involved in Lemma 2.1 from $\Theta = \Pi$ to $\Theta = p$.

<u>LEMMA 3.8</u> If the conditional distributions given H_0 resp. H_1 are absolutely continuous and $\lim_{n\to\infty} U_n = 0$ almost surely, then $\lim_{n\to\infty}\big(P_e(n+1,\,\delta(X_n))\big) - P_e(n+1,\delta(\Theta)\big) = 0$ almost surely.

<u>Remark</u> For the proof see Theorem 3 in Wolverton and Wagner (1969).

From the first assetion in Theorem 2.5 and Lemma 3.8 follows the second assertion, which completes the proof.

CHAPTER 4

ITERATION RULES WITH WEAK MEMORY

4.1 Introduction

In this chapter we are concerned with constraints under which a broad class of machine learning procedures, governed by iteration rules with memory, converge. More distinctly, we assume iteration rules with weak (or just finite) memory with respect to previous data and previously proposed estimates. Procedures of this sort offer rich possibilities to find convergent learning algorithms with powerful processing steps, and reduce in this way the time necessary for learning. We prove, in what follows, relations of interest in devising such sort of iteration rules. Use of these results, e.g., for modifying convergent learning algorithms heuristically, while retaining their convergence, and other related topics are subjects we consider in Chapter 5.

While in the previous chapters we added only some occasional extensions (on e.g., introducing truncations, admitting dependence) to well known stuffs, in the rest of these talks we mainly turn to topics pursued by the present speaker (1967, 1971, 1973a).*

(*) Added in proof: for extensions of Chapter 3 and 4 see Csibi (1973b,c).

4.2 Regulators with Weak Memory

Next let us define the properties, we assume for W, in this more general context. (We impose, in what follows, conditions (C.1) through (C.3), for any n, almost surely.)

Let

$$E\left(\left\|W_n(X^n)\right\|^2 \middle| \eta^n\right) \leq C_0 \tag{C.1}$$

for all n and almost all η^n.

We also assume weak memory with respect to previous estimates, in the sense that for some $\tau_0 > 0$

$$\left\|W_n(Z^n) - W_n\left(\overline{Z^{n-\nu,n}}\right)\right\| \leq C_1 \nu^{-\tau_0}, \tag{C.2a}$$

for all n and any non-negative ν, $\left(\overline{Z^{n-\nu,n}}\right) = \left\{\tilde{Z}_i, i \leq n\right\}$ where $\tilde{Z}_i = Z_i$ for $n-\nu < i \leq n$, and $Z_i = 0$ for $i \leq n-\nu$. I.e., $\overline{Z^{n-\nu,n}}$ is a truncation of X^n at some $\nu < n$.

If specifically, there exists some ν, for which

$$\left\|W_n(X^n) - W\left(\overline{X^{n-\nu,n}}\right)\right\| = 0 \tag{C.2b}$$

for any $\nu > \nu_0$, we say W_n exhibits, with respect to X^n, finite memory. ((C.2a) is obviously met under actual conditions, and specifically (C.2b) in digital implementation.)

We assume for the variations of $W_n(Z^n)$ the following property:

$$(C.3) \qquad \left|\tilde{V}_n\left(\overline{X^{n-\nu,n}}\right) - \tilde{V}_n\left({}_{X_n}\overline{Z^{n-\nu,n}}\right)\right| < C\delta_n ,$$

for any n, and any nonnegative integer ν. $\tilde{V}_n(Z^n) = <\tilde{X}_n, W_n(Z^n)>$
$\delta = \sup_{n-\nu<i\leq n}\|X_n - X_i\|$.
Here ${}_{X_n}\overline{Z^{n-\nu,n}} = \{Z_i, i \leq n\}$ where $Z_i = X_n$ if $n-\nu<i\leq n$ and $Z_i = 0$ for $i \leq n-\nu$.

Check in what extent and under what conditions do regulators, introduced in Chapter 2, and their moving averages meet Condition (C.3)! Obviously, Condition (C.3) may be still met within broad classes of actual problems.

In addition to these restrictions (which are loose in the sense that meeting them scarcely makes any difficulty) let us impose on the regulator W also some more significant constraints.

Obviously, the main difficulty, is in guaranteeing convergence for such classes of iteration rules with memory, the constraints of which have to be formulated step by step, at each n. (Well known memoryless variants of such classes, viz., (i) the algorithms governed by costs and (ii) the potential function type learning algorithms, have been presented in Chapter 2.)

We consider, therefore, in what follows, additional significant constraints which are specifically of interest in these cases.

<u>DEFINITION</u> Let, in this more general context, have

$$U_n = \|X_n - \Theta\|^2, \tag{4.2.1}$$

and

$$V_n = -E\left(\langle X_n - \Theta, W({}_{X_n}Z)\rangle \mid \eta^n\right), \tag{4.2.2}$$

for all n and almost all $\eta^n . ({}_{X_n}Z^n) = \{Z_i = X_n, i \le n\}$. Occasionally the value of V_n will also be compared for distinct regulators W and $\tilde{W}$. In such cases we will use the more detailed notation $V = V(W)$.

If we are only concerned with sequences of estimates generated by (1.4.12), any constraint on U and V, obviously, specifies some property of the regulator W.

In addition we assume, as counterparts to the conditions in Theorem 3.2 and 3.3, that either of the following two alternative constraints holds:

Alternative "a"

$$V_n \ge 0$$

and

$$V_{n+1} \le V_n + \alpha_n C_{10} + \xi_n$$

for all $n \ge 0$.

Alternative "b"

$$\lim_{i \to \infty} U_{n_i} = 0,$$

in probability, for any $\{n_i\}_{i=0}^{\infty}$, for which

$$\lim_{i\to\infty} V_{n_i} = 0.$$

Thus, we impose on W either the significant constraints (C.4) and (C.5) or (C.4) through (C.6).

For regulators with memory we still need to impose on $\{\alpha_n\}_{n=0}^{\infty}$ slight additional constraints. As a matter of fact, we assume $\{\alpha_n\}_{n=0}^{\infty}$ to be such that, in addition to (2.2.6), also the following hold:

$$(4.2.3) \qquad \sum_{n=0}^{\infty} \alpha_n n^{-\varepsilon} < \infty \, , \qquad \sum_{n=0}^{\infty} n^{2\tau} \alpha_n \Big(\sup_{n-\lceil n^{\eta}\rceil < i \leq n} \alpha_\nu \Big) < \infty \, ,$$

for any $\varepsilon > 0$ and some $0 < \tau < 1$. $\lceil C \rceil = 1 + \text{ent}\, C$ for any non-negative C. (4.2.3) does not impose any really relevant additional constraint. One may readily see that also (4.2.3) is met by the usual choice of $\{\alpha_n\}_{n=0}^{\infty}$ of the form: $\alpha_n = C_{52}(C_{53}+n)^{-1}$.

<u>THEOREM 4.1</u> <u>If</u> (i) $\Theta \in \mathcal{A} \subset \mathcal{X}$ (ii) for the regulator W Conditions (C.1), (C.2a), (C.3) through (C.6) hold, (ii) we generate the sequence $\{X_n\}_{n=0}^{\infty}$ of estimates by iteration, according to (1.4.12), (iv) impose on $\{\alpha_n\}_{n=0}^{\infty}$ (2.2.6) and (4.2.3), and (v) Φ is a uniformly norm reducing truncation with respect to $\mathcal{A}$, then $\lim_{n\to\infty} \| X_n - \Theta \| = 0$, almost surely.

<u>Remark 1</u> Observe that the significant constraints imposed on regulator W in Theorem 4.1 viz. (C.5) through (C.6)

restrict only those properties of W which appear also when all $Z_i\ (i \leq n)$ are identical. For any other Z^n we have no essential constraint on W. This offers a considerable freedom to adopt even heuristic ideas when devising algorithms with guaranteed convergence (see Chapter 5).

Remark 2 One may also extend a broad class of well known temporary continuous iteration procedures, in an almost similar way, to regulators with memory. (As background reading in this respect may serve such papers and monographs as those by Cypkin (1968, 1970), Driml and Nedoma (1960), Has'minskij (1969), Has'minskij and Nevel'son (1971), Sakrison (1964) and also a paper by the present speaker (1967)).

Proof The problem of proving Theorem 4.1 is how to relate the conditional expectations $E(\langle X_n - \Theta, W_n(X^n)\rangle | \eta^n)$ and $-V_n$ (the former being of interest in convergence studies, and the latter with respect to properties (C.4) (C.5a) and (C.5b), though the conditioning is, in these two cases different. One may remove this difficulty by means of (C.2) and (C.3), as we show this, precisely, in Lemma 4.2. The main point is, therefore, the proof of Lemma 4.2. Having done this, Theorem 4.1 may be readily reduced to Theorem 3.2. (We will briefly write W_n instead of $W_n(X^n)$.)

Let

$$\hat{X}_{n+1} = X_n + \alpha_n W_n. \qquad (4.2.4)$$

From this and the uniformly norm reducing property of Φ with respect to $\mathcal{A}$, obviously follows

LEMMA 4.1

$$\|X_n - \xi\| \leq \|\hat{X}_n - \xi\|$$

for any $\xi \in \mathcal{A}$.

LEMMA 4.2 For X the following inequality holds

$$E\left(\|\tilde{X}_{n+1}\|^2 \mid \eta^n\right) \leq \|\tilde{X}_n\|^2 - 2\alpha_n V_n + \tilde{S}_n$$

where $\tilde{S}_n$ is a non-negative valued sequence of random variables, measurable, at each n, with respect to X^n, such that $\sum_{n=0}^{\infty} E\tilde{S}_n < \infty$, (Recollect that $\tilde{X}_n = X_n - \Theta$.)

Remark The inequality, in Lemma 4.2 together with the constraints on the regulator W and iteration rule (1.4.12), says that, as $n \to \infty$ approaches to a supermartingale.

Proof of Lemma 4.2 From (1.4.12) and (1.5.1)

$$E\left(\|\tilde{X}_{n+1}\|^2 \mid \eta^n\right) \leq \|\tilde{X}_n\| + 2\alpha_n E\left(\langle \tilde{X}_n, W_n \rangle \mid \eta^n\right) + \tag{4.2.5}$$

$$+ \alpha_n^2 E\left(\|W_n\|^2 \mid \eta^n\right)$$

Let us consider the following approximation

$$E\left(\langle\tilde{X}_n,W_n\rangle\middle|\eta^n\right) = -V_n + \langle\tilde{X}_n, h_n(\nu)+\tilde{h}_n(\nu)\rangle + g_n(\nu) \qquad (4.2.6)$$

for all n , where

$$h_n(\nu) = E\left(W_n\middle|\eta^n\right) - E\left(W_n(X^{\overline{n-\nu,n}})\middle|\eta^n\right),$$

$$g_n(\nu) = E\left(\tilde{V}_n(X^{\overline{n-\nu,n}})\middle|\eta^n\right) - E\left(\tilde{V}_n({}_{x_n}Z^{\overline{n-\nu,n}})\middle|\eta^n\right),$$

$$\tilde{h}_n(\nu) = E\left(W_n({}_{x_n}Z^{\overline{n-\nu,n}})\middle|\eta^n\right) - E\left(W_n({}_{x_n}Z^n)\middle|\eta^n\right).$$

$\left({}_{x_n}Z^n = \left\{Z_i, i \le n\right\}\right.$ where $Z_i = X_n$.)

Let

$$q_n(\nu) = \begin{cases} h_n(\nu), \\ \tilde{h}_n(\nu), \end{cases}$$

and $\nu = \lceil n^\tau \rceil$ $(0 < \tau < 1/4.)$

From this and (C.2):

$$\left\| q_n(\lceil n^\tau \rceil)\right\| \le C_5 n^{-\tau_1}, \qquad (4.2.7)$$

almost surely for any $n > n_0 > 0$ $(\tau_1 = \tau_0\tau)$.

From the triangle inequality and the uniformly norm reducing property of Φ with respect to $\mathcal{A}$:

$$\|X_n - X_i\| \le \sum_{k=i}^{n-1}\|X_{k+1} - X_k\| \le \sum_{k=i}^{n-1}\alpha_k\|W_k\|. \qquad (4.2.8)$$

Observe that, for $Z^n = X^{\overline{n-\nu,n}}$ as well as $Z^n = {}_{X_n}Z^{\overline{n-\nu,n}}$, we have $Z_i = 0$, for $i \leq n-\nu$. The deviation between these two sequences is, therefore (in the sense given in (C.3));

(4.2.9) $$\delta_n = \max_{n-\nu<i\leq n} \|X_n - X_i\| \leq \sum_{i=n-\nu} \|X_n - X_i\|$$

From (4.2.8) and (4.2.9):

(4.2.10) $$\delta_n \leq \sum_{i=n-\nu}^{n-1} \sum_{k=i}^{n-1} \alpha_k \|W_k\| .$$

Observe that $\delta_n \geq 0$, and the property that δ_n is measurable with respect to X^n, for all n.

From (4.2.7) and Schwartz's inequality:

(4.2.11) $$|<\tilde{X}_n, h_n([n^\eta]) + \tilde{h}_n([n^\eta])>| \leq C_{66} \|\tilde{X}_n\| n^{-\tau}$$

for all $n > n_0$, almost surely.

From (4.2.6), (4.2.11), the definition of δ_n and (C.3):

(4.2.12) $$2\alpha_n E(<\tilde{X}_n, W_n> | \eta^n) \leq -2\alpha_n V_n + S_n$$

Here

(4.2.13) $$0 < S_n = 2\alpha_n(\|X_n\| C_{66} n^{-\tau} + C_{67}\delta_n)$$

From (4.2.5) and (4.2.12) :

$$E\left(\left\|\tilde{X}_{n+1}\right\|^2 \middle| \eta^n\right) \leq \left\|\tilde{X}_n\right\|^2 - 2\alpha_n V_n + \tilde{S}_n \qquad (4.4.14)$$

Here

$$\tilde{S}_n = S_n + \alpha_n^2 E(\|W_n\|^2 | \eta^n). \qquad (4.2.15)$$

From $E\|X_0\| < \infty$, (4.2.14), and (C.1):

$$E\left\|\tilde{X}_n\right\|^2 < C_0 \qquad (4.2.16)$$

From (4.2.13), (4.2.15) and (4.2.10):

$$E\tilde{S}_n \leq \alpha_n\left(C_{68} n^{-\tau_1} + C_{69} n^{2\tau} \cdot \left(\max_{n-n^\tau < i \leq n} \alpha_i\right) + C_{10}\alpha_n\right). \quad (4.2.17)$$

Since $E(\|W_n\|^2|\eta^n)$ is a non-negative valued random variable, measurable with respect to X^n , and all right side terms in (4.2.15) are, for given n , fixed positive reals, $\tilde{S}_n$ is also a non-negative valued random variable. It is, obviously, also measurable with respect to X^n .

It follows from (4.2.17), (2.2.6) and (4.2.3) that there exists such a $0 < \tau < 1$, for which

$$\sum_{n=0}^{\infty} E\tilde{S}_n < \infty,$$

by which we proved that all conditions imposed for $\tilde{S}_n$ in Lemma 4.2 hold. This completes the proof of Lemma 4.2.

Proof of Theorem 4.1 Let us adopt the following substitutions: $\nu_n = 0$, $\beta_n = \alpha_n$ and $\xi_n = \tilde{S}_n$.
Then (3.3.5) follows from Lemma 4.2, and Condition (vi) in Theorem 3.3 from (C.6). Thus, the sequences, $\{U_n\}_{n=0}^{\infty}$ and $\{V_n\}_{n=0}^{\infty}$ of random variables meet all the conditions given in Theorem 3.3, from which $\lim_{n\to\infty} \|X_n - \Theta\| = 0$, almost surely, follows. This completes the proof of Theorem 4.1.

THEOREM 4.2 If (i) $\Theta \in \mathcal{A} \subset \mathcal{X}$ (ii) the samples ω_n, $n = 0,1,\ldots$ are drawn from Ω totally independently according to a probability distribution Q, (iii) the regulator W is such that $\|W_n\| < C_{71}$ as well as $|W_n(\omega)| < C_{71}$, for all n and ω, and Conditions (C.2.a), (C.3) and (C.5) hold, (iv) we generate the sequence $\{X_n\}_{n=0}^{\infty}$ of estimates by iteration, according to (1.4.12) (v) impose on $\{\alpha_n\}_{n=0}^{\infty}$ Constraints (2.2.6) and (4.2.3), and Φ is a uniformly norm reducing truncation with respect to $\mathcal{A}$, then $\lim_{n\to\infty} V_n = 0$ almost surely.

Proof We reduce the problem to Theorem 3.4.

Adopting $U_n = \|\tilde{X}_n\|^2$ inequality (3.3.5) follows from Lemma 4.2. According to our assumptions W meets also the conditions imposed in Theorem 2.2.

4.3 On Regression Problems with Memory

Assume we may, at any instant n, evaluate the

performance of the learning procedure by some appropriate cost function $\tilde{Y}_n$ which, however, may have memory with respect to previous estimates. More distinctly assume that: $\tilde{Y}_n = \tilde{Y}_n(\omega^{n+1}, \ell^{n+1}, \tilde{\omega}^{n+1})$.

One may of course consider, as a parameter of $\tilde{Y}_n(X^n)$, the sequence X^n of all previous estimates up to n, and seek in the usual way, for such X^n for which the associated risk becomes minimal.

However, this problem is not of actual interest in many cases of machine learning. It is therefore, worthwhile to stop for a while at the question, what one really wants to do in such cases.

Observe that within a wide scope of actual learning problems, the final form of the estimate has to be developed within some finite training period, and one has actually to use the classifier only at such instants n at which the after-effects of this training period have for long disappeared; and at such instants one may well replace X^n by $_\xi Z^n$ ($\xi \in \mathcal{A}$ is some constant)

Assume that $\tilde{Y}_n(\xi)$ is conditionally stationary, at least in the sense that $E\left(\tilde{Y}_n(\xi) \middle| \eta^n\right) = R(\xi)$ is independent of n*. If $\tilde{Y}$ is such, and the classifier is really to be used only after the training has ended, we obviously have, in the course of the learning procedure, to search for such a function Θ, for which

(*) Added in proof: This is a well known usual case if the sequence η is totally independent. For further extensions to weak dependence see Csibi (1973a,b).

$R(\xi) = \min!$, if $\xi = \Theta$. (I.e. Θ is a minimum of the regression function R.)

In machine learning we may, in many cases, introduce $\tilde{Y}$ ourselves, and may, therefore, compute at any n either its gradient (if it exists) or look for some other function which may take, in finding Θ, a similar role.

Any function Y_n may be used instead of $\tilde{Y}_n$ for this purpose, provided $E(Y_n(\xi)|\eta^n)$ behaves, for all $\xi \in \mathcal{A}$, like $\text{grad}_\xi R$.

More distinctly, assume that Y is such that (i) $E(Y_n(\xi)|\eta^n)$ is independent of n, and (ii) $r(\xi) = \text{grad}_\xi R$, for any $\xi \in \mathcal{A}$. (We term any such Y_n a quasi-gradient of $\tilde{Y}_n$. Obviously, great many functions may serve as a quasigradients, given $\tilde{Y}$.)

The first of these relations means a sort of conditional stationarity. The second enables one, this time again, to introduce also heuristic ideas, when associating a quasi-gradient Y with some given $\tilde{Y}$. (Observe that in all of these cases one may find Θ among the roots of the regression function r.)

That any such quasi-gradient may replace $\text{grad}_\xi \tilde{Y}_n$, follows from

<u>THEOREM 4.3</u> <u>If</u> (i) $\Theta \in \mathcal{A} \subset \mathcal{X}$ (ii) for the specific choice of the regulator $W_n = -Y_n$ $(n = 0,1,\ldots)$ Conditions (C.1)-(C.2a), (C.3) and (C.5b) hold, (iii) we generate the sequence $\{X_n\}_{n=0}^{\infty}$ of estimates by iteration, according to (1.4.12), (iv) impose on $\{\alpha_n\}_{n=0}^{\infty}$ Constraints (2.2.6) and (4.2.3), and (v) Φ is a uniformly norm reducing truncation with respect to $\mathcal{A}$, then

$\lim_{n \to \infty} \|X_n - \Theta\| = 0$ almost surely.

Proof Take $W_n = -Y_n$, for all n. Then, and $V = V^{(o)}$ and the proof obviously follows from Theorem 4.1 and Lemma 3.7.

4.4 Extending Potential Function Type Learning Algorithms

We illustrate the use of Theorem 4.2 only for potential function type algorithms, according to Case B (Section 2.3).

THEOREM 4.4 If (i) $\Theta \in \mathcal{A} \subset H(K)$ (ii) $W = W^{(11)}$ where $W_n^{(11)} = \gamma_n K(., \omega_{n+1})$, for all n, and $\{\gamma_n\}_{n=0}^{\infty}$ is according to Case B in Section 2.3, (iii) $K(\omega, \omega) < C_{80} < \infty$ for all $\omega \in \Omega$ (iv) the samples ω_n, $n = 0,1,\ldots$ are drawn totally independently according, at each n, to the same probability distribution Q (v) we generate the sequence $\{X_n\}_{n=0}^{\infty}$ of estimates by iteration, according to (1.4.12), (vi) impose on $\{\alpha_n\}_{n=0}^{\infty}$ Constraint (2.2.6) and (vii) Φ is a uniformly norm reducing truncation with respect to some bounded $\mathcal{A}$, then $\lim_{n \to \infty} V_n = 0$, almost surely.

Proof In order to adopt Theorem 4.2, the only thing to prove is that for $W^{(11)}$ Condition (C.3) holds. (Obviously $W^{(11)}$ is memoryless and therefore (C.2) is met. On the other hand (C.1) is met because of Condition (iii) in Theorem 4.2).

Observe that, for an $\{\gamma_n\}_{n=0}^{\infty}$, according to Case B, we have

(4.4.1) $$V_n^{(2)} = V_n^{(2)}(X^n) = \int_\Omega |X_n(\omega) - \Theta(\omega)|^2 Q(d\omega)$$

(See (2.3. 9)).

Recollect Lemma A.2 (Appendix 2). By this and (4.4.1)

$$\left| \int_\Omega \left((Z_n(\omega) - \Theta(\omega))^2 - (\tilde{Z}_n - \Theta(\omega))^2 \right) Q(d\omega) \right|$$

$$= \left| \int_\Omega \left(Z_n(\omega) - \tilde{Z}_n(\omega) \right) \left(Z_n(\omega) + \tilde{Z}_n(\omega) - 2\Theta(\omega) \right) Q(d\omega) \right|$$

(4.4.2) $$\leq C_{90} \| Z_n - \tilde{Z} \|^2 \left(\| Z_n \| + \| \tilde{Z}_n \| + 2 \| \Theta \| \right)^2$$

For any bounded $\mathcal{A}$, and Z and $\tilde{Z}$ in $\mathcal{A}$: $\| Z_n \| + \| \tilde{Z}_n \| + 2\| \Theta \| < C_{91}$. Obviously for any Z and $\tilde{Z}$, for which $\delta_n = \sup_{i \leq n} \| Z_i - \tilde{Z}_i \|$.

$$| V_n^{(2)}(Z^n) - V_n^{(2)}(\tilde{Z}^n) | < C_{91} \delta_n .$$

Thus (C.3) is met. This completes the proof of Theorem 4.4.

Remark 1 By Theorem 4.4, we have also an illustration for the convergence of such potential type learning algorithms, for which $\Phi \neq I$.

Remark 2 It is easy to extend Theorem 4.4 also to regulators with memory of such forms as $W = W^{(12)}$, where

$$W_n^{(12)} = \left(\sum_{i=n-n_0}^{n+1} \gamma_i \right) K(\cdot, \omega_{n+1}) .$$

Remark 3 It is of interest to compare the possibilities, offered by introducing algorithms in this way, with results due to Györfi (1971), concerning also regulators, which are linear combinations of terms of the form $\gamma_i K(.,\omega_{n+1})$.

CHAPTER 5

EXTENDING TECHNIQUES, INCLUDING IMPLEMENTAL CONSTRAI NTS

5.1 Introduction

One has actually to solve two sort of problems when devising learning algorithms. Viz., (i) one has to obtain appropriate experimental and, if possible, also theoretical evidence that the procedure in question is really able to learn, in more or less time, the target O of learning; however (ii) one has also to make efficient use of what is a priori known about the specific properties of the problem. While heuristic ideas concerning these latter properties are usually of help in some initial period of the learning procedure, one usually needs in the final period a mathematical guarantee that the procedure will really approach to Θ .

Obviously, there are questions to be considered in both phases of the overall learning procedure. There are interesting further topics concerning end procedures, viz., the rate of convergence, acceleration techniques and stopping rules, we are not concerned with here. (For such problems see, e.g., Schmetter (1960), Wasan (1969) and Siraev (1969) in which a number of further references are also included. Concerning, specifically, convergence rate we also refer to Gulyas (1971) and a paper by the present speaker (1967)*). We are not concerned however with these topics.

In this chapter, one of the questions we are interested in, is how to combine opening and end procedures. This may, of course, be done by devising these two phases independently and then adopting some switch-over from one to the other. However it offers further possibilities (and may help also to define simple switch-over rules) if one knows under what constraints the two phases of learning may be smoothly combined.

We may derive such constraints by using results presented in Chapter 4.

Another question of actual interest is how to introduce implemental constraints into the previously described algorithmic approach. Approximation problems arise in this respect, concerning dimensional constraints and round-off. While we arrive in this way to algorithms which need at each step finite memory, it readily turns out that the memory needed may not be, as $n \to \infty$, kept under a uniform finite bound.

Implemental constraints present an interesting set of specific problems in iterative learning. While we treat, in the present context, these sort of problems as approximations, there are appealing efforts to use also a more direct approach to such question, by confining algorithmic design, a priori, to finite

(*) Concerning convergence rate see also the Notes of Course II (within the present Volume).

statistics. Intersting results, for known probability distributions, have been presented along this line Cover (1968, 1969a,b) and Hellman and Cover(1971).For round off problems,e.g.,Schmetterer (1960) and Wilkinson (1963) may serve as background texts.

5.2 Embeddings and Approximations

While up to now we could guarantee convergence only for regulators W which, specifically, meet a certain set of significant constraints, next we admit also the use of regulators which are just in some asymptotic sense related to such specific properties.

As an illustration of such sort of possibilities let us consider

THEOREM 5.1 If (i) $\Theta \in \mathcal{A} \subset \mathcal{X}$ (ii) the regulator W is such that for some $\tilde{W} = W_n = W_n + \hat{S}_n$ $n = 0,1,\ldots;$ $\sum_{n=0}^{\infty} \alpha_n E\|\hat{S}_n\| < \infty$. Conditions (C.1)-(C.4) are met and (iii) $V_n(W) = V_n(\tilde{W})$ for all n (iv) we generate the sequence $\{X_n\}_{n=0}^{\infty}$ of estimates by iteration according to (1.4.12), (v) impose on $\{\alpha_n\}_{n=0}^{\infty}$ Constraints (2.2.6) and (4.2.3), (vi) and Φ is a uniformly norm reducing truncation with respect to some bounded $\mathcal{A}$, then $\lim_{n\to\infty} \|X_n - \Theta\| = 0$ almost surely.

Remark 1 Theorem 5.1 offers broad possibilities to modify, while retaining convergence, the processing steps of simple algorithms. Obviously, a variety of heuristic ideas may also be included in this way. (For two examples see Sec. 5.3) We speak in these cases of embedding problems.

Remark 2 By Theorem 5.1 one may also bridge a number of implemental difficulties. While $\varkappa$ may be of a very high dimension, X_n may also be confined, at each step n, to a much lower dimension and, moreover, to a finite set. However, if there is no apriori evidence how Θ really looks like, one has to enlargen, as $n\to\infty$, the range of X_n. Nevertheless, if all errors, introduced in this way, may be included into $\tilde{S}_n$, one may keep implemental constraints and learning at the same time, under appropriate control.

Observe that the enlargening of the range of X_n may run out of available storage; which however causes no difficulty, if the learning arrives, before this happens, to an acceptable state. Obviously, further studies of these sort of questions are closely related to stopping rules. (We also refer to an actual case in this respect in Section 5.4.) (*)

Proof This follows almost obviously from Theorem 4.1 and Lemma 3.3.

THEOREM 5.2 If Condition (C.1) in Theorem 5.1 is replaced by $\|\tilde{W}_n\| < C_{101}$ as well as $|W_n(\omega,\omega)| < C_{102}$ for all n and $\omega\in\Omega$ however all other conditions of Theorem 5.1 hold, then $\lim_{n\to\infty} V_n = 0$.

Proof Follows from Theorem 4.2 and Lemma 3.3

(*) Added in proof: For further insight into needed storage see the estimations concerning finite step behaviour in Course II (within the present Volume).

<u>THEOREM 5.3</u> If Conditions in Theorem 2.5, and all conditions, apart (v), in Theorem 5.1 hold, and $\alpha_n = n^{-1}$ (for $n \geq 1$), then $\lim_{n\to\infty} I_n = 0$ and $\lim_{n\to\infty}(P_e\, n+1, \delta(X_n)) - P_e\, n+1, \delta(\Pi)) = 0$, almost surely.

<u>Proof</u> Follows from Theorems 5.1 and Lemma 3.3.

<u>THEOREM 5.4</u> If we have in Theorem 5.2, specifically, α_n according to (2.4.6), then in addition to the assertion of Theorem 5.2 we also have $\lim_{n\to\infty}\left(P_e(n+1, \delta(X_n)) - P_e(n+1, \delta(\Theta))\right) = 0$, almost surely.

<u>Proof</u> Follows from Theorem 5.2 and Lemma 3.3.

5.3 Hints to Applications

We present three problems for illustration. The first two of these concerned with embedding and the third with approximation.

<u>EXAMPLE 1</u> In deterministic optimization it is well known that, it may, in many cases, be appropriate to replace the well known gradient method by some more involved, usually heuristic, procedure.

One of these heuristic algorithms, is due to Nelder and Mead (1965), and is also well known as the simplex method for function minimization.

The idea of this procedure is e.g., in an Euclidean 3-space, to correct the iteration by inspecting the cost at the vertices of some triangle of the most recent three estimates X_{n-2}, X_{n-1} and X_n, simultaneously; and pick the next point, X_{n+1}, according to an a priori specified rule, far from the vertex, having maximal cost.

We do not enter here in describing this procedure. Let us denote the regulator comprising this rule by $W^{(4)}$, (Observe that, even in the case of regression problems, Conditions (C.1) through (C.4) in Sec. 4.2 may be, obviously, met by $W^{(4)}$.)

Let us define W, for all n, as $W_n = W_n^{(4)} - Y_n$ (where Y_n is a quasigradient of the problem) and assume for this (2.2.5) to hold. Provided we generate $\{X_n\}_{n=0}^{\infty}$ according to (1.4.12), impose on $\{\alpha_n\}_{n=0}^{\infty}$ Constraints (2.2.6) and (4.2.3), and adopt a Φ which is uniformly norm reducing with respect to $\mathcal{A}$, all conditions of Theorem 5.1 are met, and have $\lim_{n\to\infty} \|X_n - \Theta\| = 0$, almost surely.

<u>EXAMPLE 2</u> Let us show how one may further so called reinforcement techniques, by using some of the previous results.

Let us adopt ideas due to Saridis (1970) to Condition (C.3).

Let, for all n, $W_n = P_n W_n^{(j)}$ $(j = 0,1,\ldots)$, P_n denoting a diagonal matrix, having $\tilde{p}_n^{(j)} \cdot N^{-1}$, $j = 1,\ldots N$, as diagonal elements. $(0 \leq \tilde{p}_n^{(i)} \leq 1, \tilde{p}_0^{(i)} = N^{-1}, i = \overline{1,N})$. Let

$$\tilde{p}_{n+1}^{(i)} = \tilde{\mu}\, \tilde{p}_n^{(i)} + (1-\tilde{\mu})\, \lambda_n^{(i)}.$$

Here $0 < \tilde{\mu} < 1$ and $\lambda_n^{(i)} = \tau_n^{(i)} \Big/ \sum_{i=1}^{N} \tau_n^{(i)}$.

Let:

$$\tau_n^{(i)} = \begin{cases} 1 & \text{if} \quad \vartheta^{(i)} > K^{(i)} \\ (\vartheta_n^{(i)} + K^{(i)})/2K^{(i)} & \text{if} \quad |\vartheta^{(i)}| \leq K^{(i)} \\ 0 & \text{if} \quad \vartheta^{(i)} < K^{(i)} \end{cases}$$

for some a priori specified $K^{(i)} > 0 \; (i = \overline{1,N})$, and $\vartheta_n^{(i)} = (\bar{X}_{n+1}^{(i)} - X_n^{(i)})(\bar{X}_n^{i} - X_{n-1}^{(i)})$. $(X_n^{(i)} = \langle X_n, e_i \rangle$ and $\bar{X}_{n+1}$ stands for the estimate computed from X_n (1.4.12), and $W_n = W_n^{(j)}$.)

Assume an $\mathcal{A}$ which is bounded. Observe that (C.2) and (C.3) are met by such a regulator W_n. We may therefore use, regulators W_n of this sort in any of our embedding Theorems 5.1 through 5.3. One obtains, in this way, further illustrations in what broad extent may such reinforcement ideas be applied in machine learning. (Observe that taking $W_n = W_n^{(0)}$, $\Phi = I$ and $K^{(i)} = 0$ (for all i) we arrive at the basic ideas described in Section II by Saridis (1970). Observe that in our case $K^{(i)} > 0$, definitely; thus present Example 2 is just a supplement and not an extension of what is proved in Saridis (1970).)

EXAMPLE 3 Cover (1969) has pointed out a difficulty in using such algorithms as given by Wagner and Wolverton (1969); viz., that the complexity of W_n may not be kept under control.

In contrast to this Specht (1967) introduced kernel-type algorithms heuristically, using a series-truncation for the kernels. He as well as others, including also Mrs. Papai, Molnar and Gulyas (1971) and Molnar (1970), have demonstrated the efficiency of such procedures. One may get more insight under what constraints such algorithms acually learn, by considering a series-truncation of W_n to some $\tilde{N}(n)$-space$\left(\tilde{N}(n) \leq N\right)$, and such a sequence $\left\{\tilde{N}(n)\right\}_{n=0}^{\infty}$, by which the series truncation error S_n^* may be kept within $\sum_{n=0}^{\infty} \alpha_n E\left\|S_n^*\right\| < \infty$. (See Theorem 5.1.) The convergence of such an approximation is then guranteed by Theorem 5.3.

5.4 Round-off

For a comprehensive treatment of this topic see Sec. 3.2 in the subsequent Notes within the present volume.

APPENDICES

Appendix 1 - Proof of Lemma 2.1

<u>LEMMA 2.1</u> $\lim_{n\to\infty} E\left(\left(X_n(\omega_{n+1}) - \Pi(\omega_{n+1})\right)^2 \middle| \eta^n\right) = 0$ almost surely, then $\lim_{n\to\infty}\left(P_e(n+1,\delta(X_n)) - P_e(n+1,\delta(\Pi)\right) = 0$ almost surely.

We start with some comments and, only next to these, give the proof.

First, let us define exactly what we mean by probability of misclassification for $\delta(X_n)$.

<u>DEFINITION</u> The probability $P_e(n+1,\delta(X_n))$ of misclassification, provided we sort ω_{n+1} according to $\delta(X_n)$ is a random variable, measurable with respect to X^n, and defined by

$$\int_{B_0} P_e(n+1,\delta(X_n))\, P(d\lambda) = \int_{B_0} \left(1 - \chi_{A_n}(\omega_{n+1})\right) \Pi(\omega_{n+1}) + \chi_{A_n}(\omega_{n+1})\left(1 - \Pi(\omega_{n+1})\right) P(d\lambda),$$

(A.1)

for any $n = 0,1,\dots$ and $B_n \in F(\eta^n)$.($A_n = \{\omega : X_n(\omega) \geq 1/2\}$, $F(\eta^n)$ is the δ-algebra generated by η^n .)

<u>Remark</u> This definition is meaningful. The probabilities of misclassification, given X_n and ω_{n+1} , are averaged on the right side of (A.1), according to the total probability rule. Thus the integral of $P_e(n+1,\delta(X_n))$ over any $B_n \in F(X^n)$, ac-

tually provides what is usually meant by the average probability of misclassification with respect to some (A_n, A_n^c).

LEMMA A.1

$$P_e(n+1, \delta(X_n)) = E\Big(\big(1 - \chi_{A_n}(\omega_{n+1})\big)\Pi(\omega_{n+1}) + \\ + \chi_{A_n}(\omega_{n+1})\big(1 - \Pi(\omega_{n+1})\big)\Big|\,\eta^n\Big) \tag{A.2}$$

Proof Lemma A.1 obviously follows from (A.1). Observe that $P_e(n+1, \delta(X_n))$ as well as the conditional expectations, given η^n, are, by definition, measurable with respect to η^n.

Proof of Lemma 2.1

$$P_e(n+1,\delta(X_n)) - P_e(n+1,\delta(\Pi)) = E\big(\chi(\omega_{n+1})(1 - 2\Pi(\omega_{n+1})\,\big|\,\eta^n\big). \tag{A.3}$$

($\chi = \chi_{A_n} - \chi_{\tilde{A}}$. For $\tilde{A}$ see Sec. 1.3.)

By the definition of A_n and $\tilde{A}$, and from (A.3) we have

$$P_e(n+1, \delta(X_n)) - P_e(n+1, \delta(\Pi)) \geqslant 0 \ . \tag{A.4}$$

(In accordance with Theorem 1.1)

Similarly

$$E\big(\chi(\omega_{n+1})(1 - 2X_n(\omega_{n+1}))\,\big|\,\eta^n\big) \leqslant 0. \tag{A.5}$$

From (A.4) and (A.5), adopting Schwartz's inequality

$$P_e\big(n+1,\delta(X_n)\big) - P_e\big(n+1,\delta(\Pi)\big) \le 2E\Big(\chi(\omega_{n+1})\big(X_n(\omega_{n+1}) - \Pi(\omega_{n+1})\big)\Big|\,\eta^2\Big)$$

$$\le 2\sqrt{E\Big(X_n(\omega_{n+1}) - \Pi(\omega_{n+1})\Big)^2\Big|\,\eta^n\Big)}$$

Thus, if $\lim_{n\to\infty} E\Big(\big(X_n(\omega_{n+1}) - \Pi(\omega_{n+1})\big)^2\Big|\,\eta^n\Big) = 0$ almost surely, then $\lim_{n\to\infty}\big(P_e(n+1,\delta(X_n)) - P_e(n+1,\delta(\Pi)\big) = 0$, almost surely; which was to be proved.

Appendix 2 - A Property of Reproducing Kernel Hilbert Spaces

<u>LEMMA A.2</u> If $\xi,\eta\in H(K)$ where $H(K)$ is a RKHS of functions defined in Ω, then

$$\big|\xi(\omega) - \eta(\omega)\big| \le \|\xi-\eta\|\,K(\omega,\omega),$$

for any $\omega\in\Omega$.

<u>Proof</u> From the Reproducing Property and Schwartz's inequality

$$\big|\xi(\omega) - \eta(\omega)\big| = \langle\xi - \eta, K(.,\omega)\rangle \le \|\xi-\eta\|\,K(\omega,\omega),$$

which was to be proved.

REFERENCES (*)

Aǐzerman, M.A.; Braverman, E.M.; Rozonoer, L.I. "Metod potencial'nyh funkcii v teorii obučenija mašin", Nauka, 1970

Aǐzerman, M.A. "Remarks on two problems connected with pattern recognition" in Watanabe, M.S. (ed.) "Methodologies of Pattern Recognition", Academic Press, pp. 1-10, 1969

Aǐzerman, M.A.; Braverman, E.M.; Rozonoer, L.I. "Teoretičeskie osnovy metoda potencial'nyh funkcii v zadače ob obučenii avtomatov razdelniju vhodnyh situacii na klassy" Avtomatika i Telemehanika, No. 9., pp. 917-936, 1964a

Aǐzerman, M.A.; Braverman, E.M.; Rozonoer, L.I.,"Verojatnostnaja zadača ob obučenii avtomatov raspoznavaniju klassov i metod potencialnyh funkcii" Avtomatika i Telemehanika, No. 6., pp. 1307-1323, 1964b

Aronszajn, N. "Theory of reproducing kernel" Trans. Amer, Math. Soc., Vol. 63, pp. 337-404, 1950

Braverman, E.M.; Rozonoer, L.I. "Shodimost'slučainyh processov v teorii obučenija mašin I." Avtomatika i Telemehanika, No.1, pp. 57-77, 1969a

Braverman, E.M.; Rozonoer, L.I. "Shodimost'slučainyh processov v teorii obučenija mašin II." Avtomatika i Telemehanika, pp. 87-103, 1969b

Braverman, E.M.; Rozonoer L.I. "K stat'e 'Shodimost'slučainyh processov v teorii obučenija masin' I.", Avtomatika i Telemehanika, No. 2, p. 182, 1970

(*) For translations see the Mathematical Reviews.

Cover, T.M. "A note on the two-armed bandit problem with finite memory" Information and Control, pp. 371-377, 1968

Cover, T.M. "Hypothesis testing with finite statistics" Ann. Math. Statist., No. 3, pp. 828-835, 1969a

Cover, T.M. "Learning in pattern recognition" in Watanabe, M.S. (ed.) "Methodologies of Pattern Recognition" Academic Press, pp. 111-113, 1969b

Cover, T.M. Discussion, Wolverton, Ch.T.; Wagner, T.J. "Asymptotically optimal discriminant functions for pattern recognition" IEEE Trans. on Information Theory, Vol. IT-15, No. 2, 1969c

Csibi, S. "On continuous stochastic approximation", Proc. of the Colloquium on Information Theory, Bolyai Mathematical Society, 1967

Csibi, S. "On iteration rules with memory in machine learning", Problems of Control and Information Theory, Vol.1, N° 1 pp. 37-50, 1971 (partly appeared in the Proceedings of the Fourth Hawaii International Conference on System Sciences, Western Periodicals, pp. 205-207, 1971)

Csibi, S. "On embedding heuristics and including complexity complexity constraints into convergent learning algorithms" in Watanabe, M.S. (ed.) "Frontiers of Pattern Recognition", Academic Press 1972

Csibi, S. "Statistical learning processes", Preprint, Telec.Res. Inst., Budapest, 1973b (in Hungarian)

Csibi, S. "Learning Finite-Dimensional Projections of Decision Functions from Weakly Dependent Samples", Preprint, Telec.Res.Inst., Budapest, 1973c

Csibi, S. "Approximations in learning decision functions recursively", Trans.6th Prague Conf.on Info.Thy.,Statist. Decision Functions, Random Processes, Checkosl.Acad. Sc., Prague, 1973a

Cypkin, Ja.Z. "Adaptacija i obučenie v avtomatisčeskih sistemah" Nauka, 1968

Cypkin, Ja.Z. "Ocnovy teorii obučajuščihsja sistem" Nauka, 1970

Cypkin, Ja.Z.; Kel'mans G.K. "Rekurrentnye algoritmy samoobučenija "Techničeskaja Kibernetika", N° 5, 1967

Doob, J.L. "Stochastic Processes", Wiley, 1953

Driml, M.; Nedoma, J. "Stochastic approximation for continuous random processes", Trans. 2nd Prague Conf. on Info . Theory, Statistics, Decision Functions, and Random Processes, Czech. Acad. Sci., Prague, pp. 145-158, 1960

Dvoretzky, A. "On stochastic approximation", Proc. 3rd Berkeley Symp. on Math Statist. and Prob., Univ. of California Press, Vol. 1, pp. 39-55, 1956

Feller, W. "An Introduction to Probability Theory and Its Applications" Vol. 2, Wiley, 1966

Fu, K.S. "Sequential methods in pattern recognition and machine learning" Academic Press, 1968

Fu, K.S. "On sequential pattern recognition systems" in Watanabe, M.S. (ed.) "Methodologies of Pattern Recognition" Academic Press, pp. 159-202, 1969

Fu, K.S. "Learning control systems- Review and outlook" IEEE Trans. Automatic Control, Vol. AC-15, No. 2, pp. 210-221, 1970

Gladysev, E.G. "O stochasticeskoi approksimacii" Teor. Verojat.i ee Primen., Vol. 19, No. 2, pp. 297-300, 1965

Gulyás, O., "On extended potential function type learning algorithms and their convergence rate" Problems of Control and Information Theory, Vol. 1, No. 1 1972

Györfi, L. "Two theorems in Probability Theory and their use in proving the convergence of learning algorithms" Preprint, Telecommunication Research Institute, 1970 (in Hungarian)

Györfy, L. "Convergence of potential function type learning algorithms", Problems of Control and Information Theory, Vol. 1 (3-4), pp. 247-265, 1972

Has'minskii R.Z. "Ustočnivost sistem differencial'nyh uravnenii pri slučaĭnih vozmuscenijah ih paramterov" Nauka, 1969

Has'minskii R.Z.; Nevel'son, M.B. "O nepreryvnyh procedurah stohastičeskoĭ approksimacii" Problemy Peredači Informacii Vol. 7, No. 2., pp. 58-69, 1971

Hellman, M.E.; Cover, T.M. "Optimal learning with finite memory" Abstract, Ann. Math.Statist., Vol. 39, No. 5, pp. 1739-1794, 1968

Hellman, M.E.; Cover, T.M. " On memory saved by randomization" Proc. of the Fourth Hawaii International Conference on System Sciences, Western Periodicals, pp. 440-442, 1971

Ho, Y.C.; Agrewala, A.K. "On pattern classification algorithms. Introduction and survey" Proc. IEEE Vol. 56, No. 12, 1968

Kiefer, J.; Wolfwitz, J. "Stochastic estimation of the maximum of a regression function" Ann. Math. Statist. Vol. 23, pp.462-466, 1952

Loéve, M. "Probability Theory", D. Van Nostrand, 1955

Molnár, L. "The polynomial discriminant method of pattern recognition and its use in Meteorology" Proc. VIst Yugoslav International Conference on Information Processing, Bled, 1970

Nelder, J.A.; Mead. R. "A simplex method for function minimization" Computer Journal, No.4, pp. 308-313, 1965

Neven, J. "Bases Matematiques du Calcul des Probabilities", Masson et Cie, 1964 (in Russian - "Matematičeskic Osnovy teorii verojatnosteĭ", Izd., Mir, 1969)

Mrs. Pápai Szalai, G.; Molnár L.; Gulyas O., "Use of learning algorithms in Meteorology for the forcasting of convective acivity" Acta Cybernetica, Univ. Szeged, (in Russian, Vol.1, pp. 201-218, 1972

Parzen, E. "On estimation of a probability density function and mode" Annals.Math.Statist., Vol. 33, pp. 1065-1076, 1962

Proakis, J.G.; Miller, J.H. " An adaptive receiver for digital signaling through channels with intersymbol interference" IEEE Trans. IT-15 pp. 484-497, 1969

Robbins, H.; Monro, S. "A stochastic approximation method" Ann. Math.Statist., Vol. 22, pp. 400-407, 1951

Sakrison, D.J. "A continuous Kiefer-Wolfowitz procedure for random processes" Ann.Math.Statist. Vol. 35, pp. 590-599, 1964

Sakrison, D.J. "Stochastic approximation. A recursive method for solving regression problems" in Balakrishnan, A.V. (ed.) "Advances in Communication", Vol. 2 Academic Press, 1966

Saridis, G.N. "Learning applied to successive approximation algorithms" IEEE Trans. System Science and Cybernetics, Vol. SSC-6, No. 2, pp. 97-103, 1970

Saridis, G.N.; Nikilic, Z.J.; Fu, K.S. "Stochastic approximation algorithms for system identification, estimation and decomposition of mixtures", IEEE, Trans. System Science and Cybernetics, Vol. SSC-5, No. 1, pp. 8-15, 1969

Schmetterer, L. "Stochastic approximation" Proc. 4th Berkeley Symposium on Math. Statist. and Probability, Univ. Calif. Press, pp. 587-608, 1960

Sirjaev, A.N. "Statisticeskij posledovatel'nyj analiz", Nauka, 1969

Vasil'ev, V.I. "Raspoznajuscie sistemy", Naukova Dumka, 1969

Wasan; M.T. "Stochastic Approximation", Cambridge Univ. Press, 1969

Watanabe, M.S. (ed.) " Methodologies of Pattern Recognition" Academic Press, 1969

Wilkinson, J.H., "Rounding Errors in Algebraic Processes" Prentice-Hall, 1963

Wolverton, Ch.T.; Wagner, T.J., "Asymptotically optimal discriminant functions for pattern recognition" IEEE Trans. Information Theory, Vol. IT-15, No. 2, pp. 258-265, 1969

CONTENTS

COURSE II

LEARNING PROCESSES : FINITE STEP BEHAVIOR AND APPROXIMATIONS

PREFACE

It frequently happens in machine learning that what has to be learned is a member of some nonfinite space, however the learning process has to take because of digital computations, values within some finite range.

This is, for instance the case, if pairs of symptoms and diagnoses are presented to a machine after another, and we want the machine to estimate, e.g., either the a posteriori probability function or some Bayesian discrimination function, the range of which is non-finite. (The final purpose of doing so is usually to propose some reasonable decision rule.)

Approximations involved in such problems raise a number of interesting question in the theory of learning processes,an illustration of which is the main purpose of these talks. More distinctly, topics related to stability,approximations of kernels, and also behavior within finite number of steps areof particular interest in these Notes.

We confine ourselves to learning optimal decision rules recursively. For a more general formulation of learning problems, and some underlying results in stability theory the comparison CISM Notes *on learning processes (Course I in this Volume) may serve as background stuff, in which further bibliography is also included.*

S. Csibi

CHAPTER I

LEARNING DECISION RULES EMPIRICALLY

1.1 Optimal Decision Rules

Assume a teacher presenting a sequence $\eta_n = (\omega_n, d_n), n = \ldots -1, 0, 1, \ldots$ of pairs of symptoms and associated diagnoses afteranother. Let $\omega_n \in \Omega$ (where Ω denotes, e.g., a subset of some M-variate Euclidean space). Admit $d_n = 0,1,$, where the value taken by d_n declares whether, at instant n, some hypotheses H or its negation H^c actually holds.

Let us introduce a statistical study by assuming a probability space (Λ, F, P). ($\Lambda = X_{n=-\infty}^{\infty} \Omega_n, \Omega_n = \Omega$ for $n = 0, \pm 1, \ldots$ and $\omega \in \Omega_n$.) P is, apart from some very loose properties, a priori unknown.

We confine ourselves to stationary observations, i.e., the symptoms $\omega_n (n = 0, 1, \ldots)$, which may also be dependent, form a stationary sequence.

Assume $d_n (n = 0, \pm 1, \ldots)$ to be a random variable related to the sequence $\{\omega_n ; n = 0 \pm 1, \ldots\}$ of symptoms in a memoryless manner, in the following sense:

$$P(d_n = 1 | \omega_n, \eta^{n-1}) = P(d_n = 1 | \omega_n) = \Pi(\omega_n). \qquad (1.1.1)$$

(Here and in what follows, the superscript n means the past up to n. E.g., $\eta^n = \{\eta_\nu, \nu \leq n\}$. We call $\Pi = \{\Pi(\omega), \omega \in \Omega\}$, in accordance with usual terminology, a posteriori probability function.

Observe that broad classes of actual problems may appropriately be described even under Constraint (1.1.1). E.g., one may take as ω_n the sequence $\{\tilde{\omega}_\nu, n-\nu_0<\nu\leq n\}$ of observations made within $(n-\nu_0, n)\,(\nu\geq 0)$. As a matter of fact one arrives at (1.1.1), in this case, if the diagnosis at n is related to previous observations $\tilde{\omega}^n$ only through

$$\left\{\tilde{\omega}_n,\ n-\nu<\nu\leq n\right\}.$$

This property may frequently serve as a good description of what actually happens, as actual observations are usually uniformly weakly dependent (completely regular Rosenblatt 1956, Rozanov 1963). (Observe that one obtains, in this way a dependent sequence $\{\omega_n, n=0,\pm 1,\ldots\}$ of random variables, even if the observations $\{\tilde{\omega}_n, n=0,\pm 1,\ldots\}$ are totally independent.)

Assume the probability density $p(\omega)$ that ω_n takes the value ω to exist with respect to some known σ-finite measure, μ defined in (Ω, β). (Because of stationarity the same p may be assumed for any n.)

Accordingly, the joint probability density that takes the value ω and $d_n=1$, reads $p_1(\omega)=\Pi(\omega)\,p(\omega)$. Let $p_0(\omega)=\left(1-\Pi(\omega)\right)p(\omega)$.

Let $P_i=P(d_n=i)$ denote at any n, the a priori probability of the diagnosis $d_n=i\,(i=0,1)$. We may, under these conditions introduce any of the well known Bayesian discrimination functions, the simplest form of which is

$$D(\omega) = P_1\, p_1(\omega) - P_0\, p_0(\omega) = \big(2\Pi(\omega)-1\big)\, p(\omega). \qquad (1.1.2)$$

Let

$$\tilde{A} = \left\{\omega : \Pi(\omega) \geq \frac{1}{2}\right\} = \left\{\omega : D(\omega) \geq 0\right\}.$$

If one declares, at instant n , the hypothesis H true if $\omega \in A$ the "standard" decision rule $(\tilde{A}, \tilde{A}^c)$ is obtained (The supersctipt c stands for the complement.) For $(\tilde{A}, \tilde{A}^c)$ the probability of misclassification reads

$$P_e(\tilde{A}) = P\left[(d_0 = 1)(\omega_0 \in \tilde{A})\, \cup (d_0 = 0)(\omega_0 \in \tilde{A})\right] =$$

$$= P(\omega_0 \in \tilde{A}) + \begin{cases} \int_\Omega \chi_A(\omega)\big(1 - 2\Pi(\omega)\big)p(\omega)\, \mu(d\omega), \\ \int_\Omega \chi_A(\omega)\, D(\omega)\, \mu(d\omega) \end{cases} \qquad (1.1.3)$$

($\chi_{\tilde{A}}$ stands for the indicator of $\tilde{A}$.)

It is well known that $(\tilde{A}, \tilde{A}^c)$ is optimal in the same sense that for any decision rule (A, A^c):

$$P_e(\tilde{A}) \leq P_e(A). \qquad (1.1.4)$$

This property obviously follows from

$$P_e(A) - P_e(\tilde{A}) = \begin{cases} \int_\Omega \tilde{\chi}(\omega)\, (1 - 2\Pi(\omega)\, p(\omega)\, \mu(d\omega), \\ \int_\Omega \tilde{\chi}(\omega)\, D(\omega)\, \mu(d\omega), \end{cases} \qquad (1.1.5)$$

and the fact that

$$(1.1.6)\qquad \left.\begin{array}{l}\tilde{\chi}(\omega)\left(1 - 2\Pi(\omega)\right)\\ \tilde{\chi}(\omega)\, D(\omega)\end{array}\right\} \geq 0$$

$(\tilde{\chi} = \chi_{\tilde{A}} - \chi_A)$.

Observe that Π and D may equally well serve for devising an optimal decision rule (in the aforementioned sense of the probability of misclassification), provided p is a priori known. However, one has to make a distinction between the role of Π and D in learning problems, in which we have no such a priori knowledge.

From (1.1.5) follows that the excess probability $P_e(A) - P(\tilde{A})$ of misclassification may be computed form D even if p is unknown, while this can not be done for Π. (One may give in this latter case only rough estimates on $P_e(A) - P(\tilde{A})$ provided there is some evidence how to uniformly overbound $p(\omega)$ for all $\omega \in \Omega$.)

If $\{\omega_n\, ;\, n = 0, \pm 1 \ldots\}$ is a dependent sequence, one may also be interested in the conditional probability $P_e(A, \omega^\circ)$ of misclassifying ω_1, given ω°. (A in the argument refers to the fact that the decision rule (A, A^c) is used.) Viz.,

$$(1.1.7)\qquad P_e(A, \omega^\circ) = P\left[(d_1 = 1)\,(\omega_1 \in A^c) \cup (d_1 = 0)(\omega_1 \in A) \,\middle|\, \omega^\circ\right]$$

From (1.1.7) follows

$$(1.1.8)\qquad P_e(A, \omega^\circ) = P(\omega_1 \in A \mid \omega^\circ) + E\left(\chi_A(\omega_1)(1 - 2\Pi(\omega_1)) \mid \omega^\circ\right)$$

From (1.1.8) and (1.1.6) follows that also, in this case, $(\tilde{A},\tilde{A}^c)$ is uniformly optimal, in the sense that

$$P_e(\tilde{A},\omega^0) \le P_e(A,\omega^0), \qquad (1.1.9)$$

for any $A\subset\Omega$ and almost all ω^0.

We are primarily interested, also in this case in the excess probability of misclassification

$$P_e(A,\omega^0) - P_e(\tilde{A},\omega^0) = E\left(\chi(\omega_1)(1-2\Pi(\omega_1)\,\middle|\,\omega_0\right). \qquad (1.1.10)$$

Obviously one can not, in this case replace (1.1.10) by a relation independent of the unknown measure, as we did in (1.1.5), simply by adopting the discrimination function D unless the symptoms are independent.

In a number of applications we are interested, instead of setting up any decision rule, in a detailed description of how diagnoses and symptoms are related at each ω (for which, say $p(\omega) > \varepsilon > 0$). This topic is an example of problems in which, specifically, the a posteriori probability function Π has to be considered.

We consider in the present context Bayesian discrimination functions just for the sake of illustration. Most of what follows may also be instructive if likelihood functions have to be learned: case of interest in many actual classification problems. (see, e.g. Cyplein, 1970).

1.2 Learning Optimal Discrimination Functions

Assume Π as well as D to be square integrable with respect to μ, i.e.

$$\int_{\Omega} f(\omega^2)\,\mu(d\omega) < \infty ,$$

for $f = \Pi$ and $f = D$, respectively.

Let $\{\varphi_i\}_{i=1}^{\infty}$ be a complete orthonormal system in $\mathcal{L}_2(\mu)$, $\varkappa$ a subspace spanned by $\{\varphi\}_{i=1}^{N} \in \mathcal{L}_2(\mu) = (\varphi_i = \{\varphi_i(\omega), \omega \in \Omega\})$, and $N < \infty$. Let us denote by f^* the projection (mean square estimate) of any $f \in \mathcal{L}_2(\mu)$, with respect to $\varkappa \subset \mathcal{L}_2(\mu)$.

We are going to devise, at each $n = 0,1,\ldots$, estimates of D^*, D or Π recursively, admitting as X_n some appropriate rnadom variable which is $F(\eta^n)$-measurable. (D and D^* will be estimated only for totally independent sequences of symptoms.*)

One may have apriori evidence that D^*, or Π (whichever has to be learned) is within some subset $\mathcal{A} \in \mathcal{L}_2(\mu)$. It is then reasonable to confine $X_n (n = 0,1,\ldots)$ also to $\mathcal{A}$, as we actually do so in the sequel.

Let us consider a sequence $\{X_n ; n = 0,1,\ldots\}$ of estimates generated by an iteration rule of the following form:

$$X_{n+1} = \Phi\left(X_n + \alpha_n W_n\right). \tag{1.2.1}$$

Here $\alpha_n > 0$. W_n is, at any $n = 0,1,\ldots 0$, a random variable defined by the designer. By $W_n(X^n)$ teaching up to n is to be taken into account. Accordingly, we admit only such W_n $(n = 0,1,\ldots)$ which is $\mathcal{F}(\eta^{n+1})$-measurable. (The notation $W_n = W_n(X)$ will also make sense in the sequel, as we will also consider the pro-

(*) Added in proof: For further extensions see Csibi (1973b,c)

perties of $W_n(Z^n)$ for various sequences $Z^n=\{Z\nu, \nu \leq n\}$ of $\mathfrak{F}\{\eta^n\}$ measurable, random variables, taking values in $\mathcal{A}$ but not necessarily generated by (1.2.1).

$\Phi : \mathcal{X} \to \mathcal{A}$ is a mapping by which $\{X_n;\ n = 0,1,\ldots\}$ may be kept within $\mathcal{A}$. We admit also in the present study only such Φ which are uniformly norm reducing with respect to $\mathcal{A}$, in the sense that

$$\|\Phi(\xi)-\tilde{\xi}\| \leq \|\xi-\tilde{\xi}\| , \tag{1.2.2}$$

for any $\xi \in \mathcal{X}$ and $\tilde{\xi} \in \mathcal{A}$.

It may computationally be appropriate to define $\mathcal{A}$ and Φ in terms of coordinates with respect to $\{\varphi_i\}_{i=1}^{\infty}$. This may be done by taking

$$\mathcal{A} = \left\{ g = \sum_{i=1} c_i \varphi_i : a_i \leq c_i \leq b_i ;\ i = 1,2,\ldots \right\} \tag{1.2.3}$$

where $a_i \leq b_i \ (i = 1,2,\ldots)$ are apriori fixed reals.

One may introduce a uniformly norm reducing mapping Φ with respect to $\mathcal{A}$, by applying $\{\Phi_i\}_{i=1}$ on the coordinates $\{c_i\}_{i=1}$ of any $Z = \sum_{i=1}^{\infty} c_i \varphi_i \in \mathcal{L}_2(\mu)$ in the following way:

$$\Phi(c_i) = \begin{cases} c_i & a_i \leq c_i \leq b_i , \\ a_i \quad \text{if} & b_i < c_i , \\ b_i & c_i < a_i , \end{cases} \tag{1.2.4}$$

$(i = \overline{1,N}.)$

Let our aim be devising a sequence $\{X_n ; n = 0,1,...\}$ of estimates by which also an optimal decision rules may be learned (either precisely or, at least, approximatively). Accordingly, let us propose at each $n = 0,1,...$ the following decision rule:

Let

$$A_n = \begin{cases} \left\{\omega : X_n(\omega) \geqslant \frac{1}{2}\right\}, \\ \left\{\omega : X_n(\omega) \geqslant 0\right\}, \end{cases} \tag{1.2.5}$$

(depending on wether Π or D is estimated by X_n).

Let us declare, at any instant n , the hypothesis H true provided $\omega_n \in A_n$. The probability of misclassification for (A_n, A_n^c) is obtained simply from (1.1.3) by replacing $\tilde{A}$ by A_n.

However in the case of dependent symptoms one has to introduce a conditional probability of misclassification for $\mathcal{A}_n$ in a bit revised way. Viz., we now consider

$$P_e(A_n, \eta^n) = P\Big[(d_{n+1} = 1)(\omega_{n+1} \in A_n^c) \cup \\ \cup (d_{n+1} = 0)(\omega_{n+1} \in A_n)\Big], \tag{1.2.6}$$

as A_n depends also on d^n. (Obviously if A_n changes only slightly as n proceeds, which is the case at any advanced stage of learning, $P_e(A_n, \eta^n)$ is usually close to $P_e(A_n, \omega^n)$.

Observe that the decision rule (A, A^c) is uniformly optimal for any n and almost all η^n, also in this case. Viz.,

$$P(\tilde{A}, \eta^n) \leq P(A_n, \eta^n), \qquad (1.2.7)$$

for any n and almost all η^n, due to the same reasons we already referred to when considering (1.1.9).

Next, let us introduce sequences of kernels for learning D^*, D and Π, respectively.

Example 1-1 Let us confine ourselves to independent sequences of symptoms and finite dimensional estimates $X_n (n = 0,1,\ldots) \in \varkappa$.

(I.e., we consider estimates of the form $X_n = \sum_{i=1}^N c_i^{(n)} \varphi_i$, where $N = \dim \varkappa$.) Accordingly, let our goal be to learn the projection D^*.

By definition

$$\left\| D - Z \right\|^2 = \int_\Omega \left(D(\omega) - Z(\omega)\right)^2 \mu(d\omega) \qquad (1.2.8)$$

takes its minimal value at $Z = D^*$. ($Z = \sum_{i=1}^N c_i \varphi_i$ denotes some arbitrary member of $\varkappa$. $D^* = \sum_{i=1}^N c_i^* \varphi_i$.)

Let us take in (1.2.8) in this way, the gradient with respect to $c = \left\{c_i\right\}_{i=1}^N$. We obtain in this way as a necessary condition for D^*:

$$\int_\Omega \left(D(\omega) - D^*(\omega)\right) \varphi_i(\omega)\, \mu(d\omega) = 0 , \qquad (1.2.9)$$

for $i = \overline{1,N}$.

Observe that $D(\omega) = \left[2\Pi(\omega)-1\right]p(\omega)$ and that d_0 is a conditionally unbiased estimate of $\Pi(\omega_0)$ (in the sense of (1.1.1)). Notice also that $D^* = \sum_{i=1}^{N} c_i^* \varphi_i$ and that $\{\varphi_i\}_{i=1}^{N}$ is assumed to be an orthonormal system with respect to $\mathcal{L}_2(\mu)$. Accordingly, (1.2.9) may be rewritten as

$$(1.2.10) \qquad E\left[(2d-1)\,\varphi_i(\omega_0) - c_i^*\right] = 0,$$

$i = \overline{1,N}$. I.e. the coefficients of D^* are roots of the simultaneous regression equations (1.2.10). (Because of stationarity we could, of course in (1.2.10) equally well replace ω_0 by any ω_n.)

We could rewrite (1.2.10) also into the usual vectorial form, considering the vector $\varphi(\omega_0) = \{\varphi_i(\omega_0)\}_{i=1}^{N}$. Instead of doing so, we present here another equivalent necessary condition, for D^*. Viz.,

$$(1.2.11) \qquad E\left[(2d_0-1)\,\tilde{K}(\cdot,\omega_0) - D^*\right] = 0$$

(which immediately follows from (1.2.10), and $i = \overline{1,N}$ observing that $D^* = \sum_{i=1}^{N} c_i \varphi_i$).

(1.2.12) Here $\tilde{K}(\omega, \tilde{\omega}) = \sum_{i=1}^{N} \varphi(\omega)\, \varphi_i(\tilde{\omega})$ for any $\omega, \omega \in \Omega$.

We refer to our previous Notes, more distinctly to Section 2.2 in Ref. (Course I in the present Volume) on learn ing in terms of costs and devising Robbins Monro type algorithms.

We propose, for solving (1.2.11), the following Robbins Monro type algorithm

$$W_n = W_n^{(0)} \triangleq (2d_{n+1}-1)\,\tilde{K}(.,\omega_{n+1}) - X_n . \qquad (1.2.13)$$

It needs of course further study wether the sequence of estimates $\{X_n;\ n = 0,1,\ldots\}$ generated by (1.2.13), can really be made to learn D^*. (Algorithm (1.2.19) is, in its basic form, well known from Cypkin 1970, Cypkin and Kelmans 1970, Aĭzerman, Braverman and Rozonoer 1970.) An extension is due to Györfi (1972), and will be presented at a seminar within the present course.

Observe that (1.2.13) generates a sequence of estimates of a priori fixed finite dimension N. While estimates of this sort still need a further a round-off in the case of digital implementation, this dimensional constraint is a crucial step toward actual computations.

By adopting a mapping according to (1.2.13), the computation of $X_{n+1} = \sum_{i=1}^{N} c_i^{(n+1)} \varphi_i$ may be carried out in a simple coordinate-wise way as follows

$$c_{n+1}^{(i)} = \Phi_i\left[c_i^{(n)}(1-\alpha_n) + (2d_{n+1}-1)\varphi_i(\omega_{n+1})\right] \qquad (1.2.14)$$

Obviously, as we constrain $X_n (n = 0,1,\ldots)$ to an a priori fixed dimension, we confine ourselves to learning D^* and leave the questioh, how close D^* is to D, open for further studies (which is handicap of introducing at the outset a dimensional constraint).

It is also to be noticed that one may replace in (1.2.12) by any positive definite function K for which $K(\omega,\omega) < c_0$ and learn the projection of D onto the reproducing Kernel Hilbert space $H(K)$ generated by K. (Gulyás [1972], Györfi [1972] For the theory of reproducing Kernel Hilbert spaces see Aronszajn [1959], Kailath [1965], Meschkowski [1962] and Parzen [1962]. We replace in this case $\mathcal{H}$ by $H(K)$. (Here, and in what follows c_i, $i = 0,1,\ldots$ denotes an a priori fixed positive constant.)

As a matter of fact it is just a particularly appropriate particular choice to have $K = \tilde{K}$.

<u>Example 1.2</u>. One may learn, by observing an independent sequence of symbols, also D itself if the kernel $\tilde{K}$ or K is replaced, at any n, by some other kernel of a more general form. Let us admit step dependent kernels K_n for this purpose. We obtain in this way algorithms such as follows:

$$W_n = W_n^{(1)} \triangleq (2d_{n+1} - 1)K_n(\cdot, \omega_{n+1}) - X_n \tag{1.2.15}$$

We will consider in the sequel, specifically, the class of kernels K_n proposed by Wolverton and Wagner (1969). Viz., we assume Ω to be an M-variate Euclidean space, take $\mu(d\omega)/d\omega = 1$, and consider

$$K_n(\omega, \tilde{\omega}) = \frac{1}{h_n^M} k\left(\frac{\omega - \tilde{\omega}}{h_n}\right) \tag{1.2.16}$$

where $0 < k(\omega) < C_1$, $\int_{\Omega} k(\omega)\, d\omega = 1$

$$h_{n+1} > h_n > 0 \ (n = 0,1,\ldots), \ \lim h_n \to 0.$$

We will, of course, also impose further constraints on $k(.)$ as well as $h = \{h_n, n = 0,1,\ldots\}$ in the sequel to assure the stability of the corresponding learning process $\{X_n;\ n = 0,1,\ldots\}$

This kind of algorithms represent an iterative version of the well known kernel type estimates due to Rosenblatt (1956) and Parzen (1962), and proposed originally for probability density estimations. (Further properties of algorithms generated by (1.2.15) has been studied recently in further details by Révész and Rejtö (1972).)

While algorithms generated by (1.2.15) do not include, by definition, any dimensional constraint and may therefore serve, at least in principle, for estimating precisely D itself, actually computational constraints have to be introduced also in this case in the course of the implementation. Interesting approximation problems appear in this way, some basic principles of which we discuss in the sequel.

<u>Example 1.3</u>. One may learn, even in the case of weakly dependent symbols, Π by adopting

$$W_n = W_n^{(2)} \triangleq \left(X_n(\omega_{n+1}) - d_{n+1}\right) K(.\,,\, \omega_{n+1}) \quad . \qquad (1.2.17)$$

Here K stands for any positive definite function

such that $K(\omega,\omega) < c_1$ for all $\omega\in\Omega$. If specifically $K = \tilde{K}$, we may arrive also at this kernel by the minimizing.

$$\int_{\Lambda} \left(\Pi(\omega_0) - Z(\omega_0)\right)^2 P(d\lambda)$$

and following considerations similar to those in Example 1.1.

However the minimum we seek for in this case is not Π^* but a mean square estimate within χ with respect to the probability distribution P.

Because of this additional difficulty we do not consider here this minimization problem and focus our attention, at the outset, to an arbitrary positive definite K and to the learning of $\Pi\in\chi(K)$ itself*.

We consider constraints, by which Examples 1.1 through 1.3 may be made to learn optimal or near-optimal decision rules, in Chapter 2.

(*) Learning the projection of Π with respect to P is a topic treated e.g., in Cypkin (1970) and Csibi (1973b,c)

CHAPTER 2

FINITE STEP BEHAVIOR OF SIMPLE ALGORITHMS

2.1 Some Basic Relations

Let us denote by f the goal of the learning procedure (i.e. the function we wish to estimate). Accordingly, let

$$f = \begin{cases} D^* & 1.1 \\ D \quad \text{in Example} & 1.2 \\ \Pi & 1.3 \end{cases} \tag{2.1.1}$$

We will reduce the study of Examples 1.1 through 1.3 to a set of basic stability theorems, a brief Summary of which we give in this section.

From (1.2.1) and follows (1.2.2)

$$\left\| X_{n+1} - f \right\| \leq \left\| X_n - f + \alpha_n W_n(X^n) \right\| \tag{2.1.2}$$

For the sake of brevity, let

$$U_n = \left\| X_n - f \right\|^2$$

The properties of the sequence $\{U_n\}_{n=0}^{\infty}$ obviously depend on the way $W = \{W_n\}$ is defined, and particularly on how W and f are related. In Sections 2.2 and 2.3 we will introduce appropriate conditions for $W = W^{(j)}$ $(j = 0,1,2)$ and α, under which one of the following two supermartingale-like inequalities holds:

(2.1.3) $$E\left(U_{n+1}\middle|\eta^n\right) \le (1-\alpha_n)^2 U_n + \xi_n^{(0)}$$

(2.1.4) $$E\left(U_{n+1}\middle|\eta^n\right) \le \left(1 - C_2\alpha_n + \xi^{(2)}\right) U_n + \xi_n^{(1)}$$

for any $n \ge 0$.

Here $\xi_n^{(j)}$ are, for any $n = 0,1,\ldots$ and $j = 0,1$, non-negative valued and $\mathcal{F}(\eta^n)$-measurable, and $\sum_{n=0}^{\infty} E\xi_n^{(j)} < \infty$. $\xi_n^{(2)}$ is an a priori fixed positive valued sequence such that $\sum_{n=0}^{\infty} E\,\xi_n^{(2)} < \infty$.

One immediately obtains stability conditions in this way, as for any such sequence $\{U_n\}_{n=0}^{\infty}$ we have

<u>THEOREM 2.1</u> If $\{U_n\}_{n=\infty}^{\infty}$ is a non-negative valued sequence of $\mathcal{F}(\eta^n)$-measurable random variables for which (i) $EU_0 < \infty$ and (ii) either (2.1.3) or (2.1.4) holds, (iii) $\alpha = \{\alpha_n\}_{n=\infty}^{\infty}$ is such that

(2.1.4a) $$\sum_{n=0}^{\infty} \alpha_n = \infty \quad \text{and} \quad \sum_{n=0}^{\infty} \alpha^2 < \infty$$

<u>then</u>

$$P\left(\lim_{n\to\infty} U_n = 0\right) = 1$$

<u>Remark</u> Theorem 2.1. follows from Theorem II (Aĭzerman, Braverman Rozonoer 1969) also reproduced as Theorem 3.3 in Course I within this Volume). (Convergence specifically for (2.1.3) and $\zeta_n^{(0)} = C_5(1+n)^{-2}$ is directly proved in

Györfi [1972] in a paper to be presented also at a seminar with in this course.

One may also guarantee useful properties for such $\{U_n\}_{n=0}^{\infty}$ for finite training periods. A particularly simple and useful bound may be given for EU_n and $\alpha_n=(n+1)^{-1}$, by adopting Equation (70) in Györfi (1972) to (2.1.3), as

THEOREM 2.2 If U_n is a non-negative valued sequence of $\mathfrak{F}(\eta^n)$-measurable random variables such that (i) $EU_0 < \infty$ (ii) (2.1.3) holds, (iii) $\zeta_n^{(0)}=C_5(1+n)^{-2}$, and we adopt, specifically, (iv) $\alpha_n=(n+1)^{-1}$ then

$$EU_n \leq \frac{C_6}{n} \tag{2.1.5}$$

Proof One may readily adopt considerations in Györfi (1972) also to (2.1.3). Viz., it follows from (2.1.3) that

$$EU_{n+1} \leq EU_0 \prod_{k=0}^{n} (1-\alpha_n)^2 + \sum_{k=0}^{n-1} E\zeta_k^{(0)} \prod_{\ell=k+1}^{n} (1-\alpha_\ell)^2 + E\zeta_n^{(0)} \tag{2.1.6}$$

From condition (iii) follows

$$(1-\alpha_n) = \frac{n}{n+1} \tag{2.1.7}$$

From (2.1.6) and (2.1.7)

$$EU_{n+1} \leq \frac{1}{n+1} \sum_{k=0}^{n} E\zeta_k^{(0)} \frac{(1+k)^2}{n+1} \leq \frac{C_6}{n+1}, \tag{2.1.8}$$

where $\sum_{n=0}^{\infty} E\,\xi_k^{(0)} = C_6 < \infty$ .This completes the proof.

In order to obtain for finite training periods a similar results also for (2.1.4), we have to slightly modify the constraints imposed on $\left\{\xi_n^{(1)}\right\}_{n=0}^{\infty}$. More distinctly we have for this case (Aĭzerman Braverman, Rozonoer 1970)

<u>THEOREM 2.3</u> If $\left\{U_n\right\}_{n=0}^{\infty}$ is a non-negative valued sequence of $\mathcal{F}(\eta^n)$ -measurable random variables such that (i) $EU_0<\infty$(ii) (2.1.4) holds specifically for a $\xi_n^{(2)}$ for which

(2.1.9) $$C_{200}\,\alpha_n = \inf_{i \geq n}\left(C_2\alpha_i - \zeta_i^{(2)}\right) > 0$$

for $n \geq 1$and (iii) and $\beta = \left\{\beta_n\right\}_{n=0}^{\infty}$ are non-negative coefficients such that

$$\beta_{n+1} < \beta_n \,,\; 0 \leq \frac{(1 - C_{300}\,\alpha_n)\,\beta_n}{\beta_{n+1}} \leq 1, \quad \sum_{n=0}^{\infty} \frac{E\xi_n^{(1)}}{\beta_{n+1}} = C_3 < \infty$$

(2.1.10)

for any $n \geq n_1$ <u>then</u>

(2.1.11) $$EU_n \leq C_5\beta_n$$

for $n \geq n_1$.

<u>Proof</u> We reproduce considerations due to Braverman and Pjatnickij (1964) for the slightly broader conditions in Theorem 2.3. Let $Z_n = EU_n/\beta_n$, and observe (2.1.9). From this and

(2.1.4) follows

$$Z_{n+1} \leq Z_n + \frac{E\,\xi_n^{(1)}}{\beta_{n+1}} \quad , \tag{2.1.12}$$

for any $n \geq n \geq n_1$. From (2.1.12) and Condition (i)

$$Z_n \leq C_4 \tag{2.1.13}$$

for $n \geq n_1$.

For the constant C_4 we have

$$C_4 = \frac{EU_{n_0}}{\beta_{n_0}} + \sum_{i=n_0}^{n} \frac{E\xi_i^{(1)}}{\beta_{i+1}} \leq \frac{EU_0 \prod_{k=0}^{n_0}(1-C_{300}\,\alpha_k) + \sum_{k=0}^{n_0-1} E\xi_k^{(1)} \prod_{\ell=k+1}^{n_0}(1-C_{300}\alpha_\ell) + E\zeta_{k_0}^{(1)}}{\beta_{n_0}}$$

$$+ \sum_{i=n_0}^{n} \frac{E\xi_i^{(1)}}{\beta_{i+1}} \leq C_5 \,. \tag{2.1.14}$$

From (2.1.13) and (2.1.14) follows (2.1.11). This completes the proof.

<u>Remark</u> Observe that considerations similar to those used in proving Theorem 2.1 lead, instead of (2.1.8), to a coefficient

$$\sum_{k=0}^{n} E\xi_k^{(1)}(k+1)$$

which, however can not be kept finite, as $n \to \infty$, even in the well known case $E\xi_k^{(1)} \le C_6(n+1)^{-2}$. (See the simple case $\zeta_k^{(1)} \le \alpha_n^2 C_{61}$ and $\alpha_n = (n+1)^{-1} C_{62}$.)

However, taking $\alpha_n = \left[C_{300}(n+1)\right]^{-1}$

we obtain, from (2.1.10), $\dfrac{n}{n+1} \le \dfrac{\beta_{n+1}}{\beta_n}$,

which condition may obviously be met by $\beta_n = C_\gamma n^{-\mu}$ for any $0 < \mu < 1$. (Observe that this choice of β_n, $n = 0,1,\ldots$ is also acceptable, concerning

$$\sum_{n=0}^{\infty} \frac{E\xi_n^{(1)}}{\beta_{n+1}} < \infty .$$

See, e.g., $E\xi_n^{(1)} \le C_6(n+1)^{-2}$.

For further discussion of β see Braverman and Pjatnickij (1964).

2.2 Independent Observations

Let us show that $W^{(0)}$ and $W^{(1)}$ may actually be used for learning if the symptoms are totally independent.

By adopting $W^{(0)}$ one may learn D^* in the sense of

<u>THEOREM 2.4.</u> If $D \in \mathcal{A} \subset \mathcal{L}_2(\mu)$ and $E\|X_0\|^2 < \infty$ (ii) $\{\omega_n\}_{n=-\infty}^{\infty}$ is totally independent (iii) $\{X_n\}_{n=0}^{\infty}$ is generated by (1.2.1) adopting, specifically $W = W^{(0)}$ (iv) α is according to (2.1.4a)

(v) Φ is uniformly norm reducing with respect to $\mathcal{A}$ and (vi) we propose at $n+1$ classifying any next $\omega_\nu, \nu \geq n+1$ by the decision rule (A_n, A_n^c), <u>then</u>

$$P\left(\lim_{n\to\infty}\left\| X_n - D^*\right\| = 0\right) = 1 \tag{2.2.1}$$

and

$$P\left[\overline{\lim_{n\to\infty}}\left(P_e(A_n) - P_e(\tilde{A})\right) \leq \left\| D - D^*\right\|\right] = 1 \quad . \tag{2.2.2}$$

<u>Remark</u> Theorem 2.4. holds also for any positive definite function K, for which $K(\omega\,\omega) < C_8$. It is in this more general from proved in Györfi (1962) for $\Phi = I$. (While we admit here any $\Phi \neq I$ this mapping may, by the assumed uniformly norm reducing property, immediately be removed from further considerations).

As a matter of fact from (1.2.1), the definition of $W^{(o)}$ and the uniformly norm reducing property of Φ (with respect to $\mathcal{A}$) follows:

$$\left\| X_{n+1} - D^*\right\| \leq \left\| (X_n - D^*)(1 - \alpha_n) + \alpha_n\left[(d_{n+1} - 1)\tilde{K}(.,\omega_{n+1}) - D^*\right]\right\| \tag{2.2.3}$$

as $D^* \in \mathcal{A}$

For proving (2.2.1) one has only to show that, by squaring (2.2.3) and taking the conditional expectation (given η^n), one arrives at

$$(2.2.4) \qquad E\left(U_{n+1}\,\middle|\,\eta^n\right) \le U_n\left(1-\alpha_n\right)^2 + \xi_n^{(0)}$$

with the properties assumed for $\xi_n^{(0)}$ in (2.1.3). ($\sum_{n=0}^{\infty} E\xi_n^{(0)} < \infty$ follows from (i) $\sum_{n=}^{\infty}\alpha_n^2 < \infty$, the fact that (ii) d_{n+1} is a conditionally unbiased estimateof $\Pi(\omega_{n+1})$,that (iii) $D(\omega) = \left[2\Pi(\omega)-1\right]p(\omega)$, for any $\omega\in\Omega$and (iv) the total independence of $\left\{\omega_n\right\}_{n=\infty}^{\infty}$. For the proof of (2.2.1) see Györfi (1972).

(2.2.2) follows from (2.2.1) and

<u>LEMMA 2.1.</u>

$$(2.2.5) \qquad P_e(A_n) - P_e(\tilde{A}) \le C_9 \left\| X_n - D \right\|$$

<u>Remark</u> From (2.2.5) and (2.2.1) obviously follows (2.2.2). One has only to observe that

$$(2.2.6) \qquad \left\| X_n - D \right\|^2 = \left\| X_n - D^* \right\|^2 + \left\| D^* - D \right\|^2 .$$

<u>Proof of Lemma 2.1.</u>

Observe that similarly to (1.1.6),

$$(2.2.7) \qquad \chi_n(\omega)\; X_n(\omega) \le 0$$

($\chi_n = -\tilde{\chi}_n = \chi_{\tilde{A}} - \chi_{A_n}$. From this and (1.1.5))

$$(2.2.8) \qquad P_e(A_n) - P_e(A) \le \int_\Omega \chi(\omega)\left(D(\omega) - X_n(\omega)\right)\mu(d\omega)$$

(2.2.5) follows from (2.2.8) and Schwartz's Inequality. (Observe that $|\chi(\omega)| \leq 1$, for all $\omega \in \Omega$, and μ is assumed to be σ-finite, thus $\mu(\Omega) = C_9 < \infty$.) This completes the proof.

For the behavior of this kind of learning processes within finite steps we have, specifically for $\alpha_n = (1+n)^{-1}$,

THEOREM 2.5. If we replace Condition (iv) in Theorem 2.4 by $\alpha_n = (1+n)^{-1}$ however all other Conditions, imposed in Theorem 2.4, hold then

$$E\|X_n - D^*\|^2 \leq \frac{C_6}{n} \tag{2.2.9}$$

and

$$E\left[P_e(A_n) - P_e(A)\right]^2 \leq \frac{C_{10}}{n} + \|D - D^*\|^2 \tag{2.2.10}$$

for all $n \geq 0$. $(C_{10} = C_6 C_9.)$

Remark Theorem 2.5 follows from (2.2.4), Theorem 2.2 from (2.2.6) and (2.2.8).

While the choice of $W^{(0)}$ assures that all estimates are of the same finite dimension M, learning the projection leaves open the problem, how close D^* is to D. (While one may get some insight also in this respect by studying the specific features of the problem and making careful case studies, this question is outside of the scope of the present analysis.)

One may of course, use $W^{(0)}$ without having such a

dilemma provided (i) Ω is a finite space (as it actually appears in digital computations) and, what is more important, (ii) the dimension of $\mathcal{X}$, we obtain in this way, may still be handled within reasonable computational capabilities.

Next we show how to devise learning processes, at least principally, for learning any $D \in \mathcal{L}_2$ itself.

Let us confine ourselves to an Ω which is an M-variate Euclidean space and to $d\mu/d\omega = 1$.

Let, in addition to conditions in Example 1.2 (see 1.2)

(2.2.11)
$$\int_{\Omega} \|\omega\| \, k(\omega) \, d\omega = C_{11} < \infty$$

and

$$h_{n+1} < h_n \,, \quad \lim_{n \to \infty} \frac{h_n^M}{\alpha_n} = \infty \,, \quad \sum_{n=0}^{\infty} h_n \alpha_n < \infty \,, \quad \sum_{n=0}^{\infty} \frac{\alpha_n^2}{h_n^{2M}} < \infty$$
(2.2.12)

Under these conditions any $D \in \mathcal{A} \subset \mathcal{L}_2$ may be learned by $W^{(1)}$ (as defined in (1.2.15)) in the sense of

<u>THEOREM 2.6</u> <u>If</u> (i) $D \in \mathcal{A} \subset \mathcal{L}_2$ and $E\|X_0\|^2 < \infty$, (ii) $\{\omega_n\}_{n=0}^{\infty}$ is totally independent (iii) $\{X_n\}_{n=0}^{\infty}$ is generated by (1.2.1), adopting specifically $W = W^{(1)}$ according to (1.2.15) (iv) for $k(.)$, α and h in addition to Conditions in Example 1.2 (see 1.2), (2.2.11) and (2.2.12) hold, (v) Φ is uniformly norm reducing with respect to $\mathcal{A}$, and (vi) we propose at $n+1$, for classifying any next (A_n, A_n^c) the decision rule

then

$$P\left(\lim_{n\to\infty} \|X_n - D\| = 0\right) = 1, \tag{2.2.13}$$

and

$$P\left[\lim_{n\to\infty}\left(P_e(A_n) - P_e(\tilde{A})\right) = 0\right] = 1 \tag{2.2.14}$$

Proof From (1.2.1), the definition of $W^{(1)}$ and the uniformly norm reducing property of Φ follows

$$\|X_{n+1} - D\| \le \left\|(X_n - D)(1 - \alpha_n) + \alpha_n\left[K_n(.,\omega_{n+1})(2d_{n+1} - 1) - D\right]\right\| \tag{2.2.15}$$

From this we obtain

$$E\left(U_{n+1}\middle|\eta^n\right) \le U_n(1-\alpha_n)^2 + \xi_n^{(1)}. \tag{2.2.16}$$

Here

$$\xi_n^{(1)} = 2(1 - \alpha_n)\alpha_n E\left(< X_n - D_n, K_n(.,\omega_n)(2d_{n+1} - 1) - D >\middle|\eta^n\right) + \\ + \alpha_n^2 E\left(\|K_n(.,\omega_{n+1})(2d_{n+1} - 1) - D\|^2\middle|\eta^n\right) \quad . \tag{2.2.17}$$

Let us consider

LEMMA 2.2. Assume a $g \in \mathcal{L}_2$ such that

$$|g(\omega) - g(\tilde{\omega})| \le C_{12}\|\omega - \tilde{\omega}\| , \tag{2.2.18}$$

for all $\omega,\omega\in\Omega$. Let $k(.)$ satisfy (1.2.16) and (2.2.11). Define

(2.2.19) $$g_n(\tilde{\omega}) = \int_\Omega K_n(\omega,\tilde{\omega})\, g(\omega)\, d\omega$$

Then

(2.2.20) $$\left| g_n(\omega) - g(\omega) \right| \le C_{12} C_{11} h_n .$$

Proof of Lemma 2.2.

$$\left| g_n(\omega) - g(\omega) \right| = \left| \int_\Omega K_n(\omega,\tilde{\omega}) \left[g(\tilde{\omega}) - g(\tilde{\omega}) \right] d\tilde{\omega} \right|$$

$$\le \int K_n(\omega,\tilde{\omega}) \left| g(\tilde{\omega}) - g(\omega) \right| d\omega$$

$$= \int k(\delta) \left| g(\omega - \delta h_n) - g(\omega) \right| d\delta$$

$$\le C_{12} \int k(\delta) \left\| \delta h_n \right\| d\delta$$

(2.2.21) $$= C_{12}\, C_{11}\, h_n ,$$

by which Lemma 2.2. is proved (Here $\delta\in\Omega$. Ω is assumed to be an Euclidean M-space).

From (2.2.17) obviously follows that $\zeta_n^{(1)}$ is $F(\eta^n)$ -measurable.

Consider Lemma 2.2., the conditions imposed on $k(.)$, α and h, the fact that d_{n+1} is a conditionally unbiased estimate of $\Pi(\omega_{n+1})$. Observe also the total independence of $\{\omega_n\}_{n=0}^{\infty}$ and

$$\left(2\Pi(\omega) - 1\right) p(\omega) = D(\omega) \text{ for all } \omega\in\Omega .$$

From these facts and (2.2.17) follows $\xi_n^{(1)}>0$, for

all $n \geq 0$, and

$$\sum_{n=0}^{\infty} E\, \xi_n^{(1)} < \infty$$

Thus, for $W_n^{(1)}$ according to (1.2.15),(2.1.3) holds. As conditions in Theorem 2.6 assure that Theorem 2.1 also holds,

$$P\left(\lim_{n \to \infty} \| X_n - D \| = 0\right) = 1.$$

By this Assertion (2.2.13) in Theorem (2.6) is proved.

Assertion (2.2.14) in Theorem 2.6 follows from (2.2.13) and Lemma 2.1.

For estimations concerning the finite step behavior of learning processes defined by $W = W^{(1)}$(1.2.15) we refer to Theorem 2.3, Wolverton and Wagner (1969), Résén and Réjtö (1972).

2.3 Weakly Dependent Observations

Let us now turn to dependent observations. While mathematically we will not confine ourselves to any sort of weak dependence, the specific conditions to be introduced in the sequel will really be useful if the dependence within $\{\omega_n\}_{n=-\infty}^{\infty}$ is not too strong.

According to considerations in Section 1.2, we devote this Section to learning the a posteriori probability function Π .More distinctly, we are concerned with learning Π by means of the kernel-sequence $W^{(2)}$ (according to (1.2.17)).

As we already pointed out in Sec. 1.2., $W^{(2)}$ may be introduced by setting up a least mean square approximation problem in $\mathcal{L}_2(P)$ Viz., if $K=\tilde{K}$, one may seek for the least mean square estimate within a finite dimensional subspace defined in $\mathcal{L}_2(P)$.

While there are also known results, explaining under what conditions may such kind of algorithms tend really to the mean square error (Braverman, Pjatnickij [1964] ; Aizerman, Braverman, Rozonoer [1970]), here we do not enter into these kind of approximation studies*. Instead of this we consider in the present context just such learning problems in which may precisely be learned by $W^{(2)}$, viz. in which no complexity con-

---

(*) We are concerned also with learning the projection of Π in $\mathcal{L}_2(P)$ in Csibi (1973b,c)

As in such unconstrained approaches we do not much care to keep the kernel simple at the outset (leaving complexity constraints for further approximations) we consider here, instead of the finite dimensional kernel

$$\tilde{K}(\omega, \tilde{\omega}) = \sum_{i=1}^{N} \varphi_i(\omega)\, \varphi_i(\tilde{\omega}),$$

some more general type of kernels of essentially the same mathematical properties.

As we already mentioned in Section 1.2. any positive definite function K may serve for this purpose, provided Π may be represented by a series of $\{K(., \omega_i)\}_i$

What we need, in studying such algorithms, is such a kernel which exhibits, within some Hilbert Space $H(K)$, the following REPRODUCING PROPERTY

$$< Z, K(., \omega) > = Z(\omega) \qquad (2.3.1)$$

for any $Z \in H(K)$

We have already experienced this Reproducing Property for $\tilde{K} = \left\{\tilde{K}(\omega,\tilde{\omega}) = \sum_{i=1}^{N} \varphi_i(\omega)\varphi_i(\tilde{\omega}); \omega, \tilde{\omega} \in \Omega\right\}$ which is a positive definite function in an $\mathcal{X}$, spanned by $\{\varphi_i\}_{i=1}^{N}$.

We use the notation $H(K)$ as it is well known in the theory of Reproducing Kernel Hilbert Spaces (RKHS) (Aronszajn [1950], Kailath [1965], Meschkowski [1962], Parzen [1962], see also Gulyas [1972] and a brief remark in Course I (in the present Volume) that one and only one RKHS belongs to any po sitive definite K.

We will assume, in the sequel, a K for which $\Pi \in H(K)$. (This means that Π may be expressed in the form $\sum_{i=0}^{\infty} C_i K(.,\omega_i)$,

Next to these introductory remarks, let us prove that any $\Pi \in H(K)$ may really be learned by $W^{(2)}$ provided appropriate conditions are met.

<u>THEOREM 2.8</u> If $\Pi \in H(K)$ and $E\|X_0\|^2 < \infty$ (ii) P is such that, for any $\mathfrak{F}(\eta^n)$-measurable random variable $Z_n \in H(K)$, $C_{13}\|Z_n\|^2 \leq E\left(Z_n(\omega_{n+1})^2 \middle| \eta^n\right) \leq C_{14}\|Z_n\|^2$ (iii) $\left\{X_n\right\}_{n=0}^{\infty}$ is generated by (1.2.1), adopting specifically $W^{(2)}$ (according to (1.2.17)) and a positive definite K, for which $K(\omega,\omega) < C_{14} < \infty$ (iv) α is according to (2.1.4a), (v) Φ is uniformly norm reducing with respect to $\mathcal{A}$, and (vi) we propose at $n+1$, for classifying ω_{n+1}, decision rule (A_n, A_n^c) <u>then</u>

(2.3.2) $$P\left(\lim_{n\to\infty} \| X_n - \Pi \| = 0\right) = 1,$$

(2.3.3) $$P\left(\lim_{n\to\infty} \left(P_e(A_n) - P(\tilde{A})\right) = 0\right) = 1$$

and also

(2.3.4) $$P\left[\lim_{n\to\infty}\left(P_e(A_n, \eta^n) - P_e(\tilde{A}, \eta^n)\right) = 0\right] = 1.$$

Remark Condition (ii) may be met if $\left\{\omega_n\right\}_{n=-\infty}^{\infty}$ is, say, n_0 -dependent, the conditional probability density $p(\omega|\eta^{\overline{n-n_0,n}})$ that ω_{n+1} takes the value ω (given $\eta^{\overline{n-n_0,n}}$) exists, and $C_{14} < p(\omega|\eta^{\overline{n-n_0,n}}) < C_{14}$ for all $\eta^{\overline{n-n_0,n}}$. (Here $\eta^{n-n_0,n} = \{\eta_\nu, \nu \in (n-n_0, n]\}$.) However, observe that Condition (ii) leads to loose bounds if $p(\omega \mid \eta^{\overline{n-n_0,n}})$, exhibits extremely low and high values, as obviously happens if there is a strong dependence within $\eta^{\overline{n-n_0,n}}$ (*).

Proof From (1.2.1), the definition of $W^{(2)}$ and the uniformly norm reducing property

$$E\left(U_{n+1}\middle|\eta^n\right) \leq U_n + 2\alpha_n E\left(\langle X_n - \Pi, W_n^{(2)}(X^n) \rangle \middle| \eta^n\right) +$$

$$+ \alpha_n^2 E\left(\| W_n^{(2)}(X^n) \|^2 \middle| X^n\right). \qquad (2.3.5)$$

Because of the reproducing property and $\Pi \in \varkappa(K)$,

$$\langle X_n - \Pi, K(.,\omega_{n+1}) \rangle = X_n(\omega_{n+1}) - \Pi(\omega_{n+1}). \qquad (2.3.6)$$

Observe that d_{n+1} is a conditionally unbiased estimate of $\Pi(\omega_{n+1})$. From this and (1.2.17) we get

$$E\left(\langle X_n - \Pi, W_n^{(2)}(X^n)\rangle \middle| \eta^n\right) = -E\left(\left(X_n(\omega_{n+1}) - \Pi(\omega_{n+1})\right)^2 \middle| \eta^n\right) \quad (2.3.7)$$

By (2.3.7) and condition (ii),

(*) Added in proof. The author has in the meantime also extended approach in Chap. 4 (Course I in this Volume) to learning from weakly dependent samples and related approximation problems. (See Csibi 1973b,c).

(2.3.8) $$E\left(\langle X_n - \Pi, W_n^{(2)}(X^n)\rangle \middle| \eta^n\right) \geq - C_{13}\|X_n - \Pi\|^2$$

From this, (2.1.4a) and $K(\omega',\omega) \leq C_{14}$ follows

(2.3.9) $$E\left(U_{n+1} \middle| \eta^2\right) \leq U_n(1-\alpha_n C_{13}) + \xi_n^{(1)},$$

where

(2.3.10) $$\xi_n^{(1)} = \alpha_n^2 C_{14}$$

Thus (2.1.4) holds for $W^{(2)}$. As also all other Conditions in Theorem 2.1. are met, (2.3.2) holds.

In order to prove (2.3.3) and (2.3.4) we observe again Condition (ii). Thus

(2.3.11) $$\int_\Omega \left(X_n(\omega) - \Pi(\omega)\right)^2 p(\omega)\,d\omega \leq C_{15}\|X_n - \Pi\|^2,$$

(2.3.12) $$E\left(\left(X_n(\omega_{n+1}) - \Pi(\omega_{n+1})\right)^2 \middle| \eta^n\right) \leq C_{15}\|X_n - \Pi\|^2.$$

<u>LEMMA 2.3</u>

(2.3.13) $$\left[\left(P_e(A_n,\eta^n) - P_e(\tilde{A},\eta^n)\right]^2 \leq E\left(\left(X_n(\omega_{n+1}) - \Pi(\omega_{n+1})\right)^2 \middle| \eta^n\right)$$

<u>Remark</u> Lemma 2.3 follows from (1.1.10) in the same way as (2.2.5) from (1.1.5)*.

From (2.3.11) and (1.1.5) follows (2.3.3). From (2.3.12) and (1.1.10) follows (2.3.3). This completes the proof of Theorem 2.8.

(*) It is instructive to observe the difference between the classification tasks corresponding to them, 2.8 and (1.1.10)!

<u>THEOREM 2.9.</u> If Condition (iv) in Theorem 2.8. is replaced by Condition (2.1.8), however other Conditions in Theorem 2.8 are left unchanged then

$$EU_n < C_5 \beta_n \tag{2.3.14}$$

$$E\left(P_e(A_n) - P_e(A)\right)^2 \leq C_{16} \beta_n \tag{2.3.15}$$

and

$$\left[\left(P_e(A_n, \eta^n) - P_e(\tilde{A}, \eta^n)\right]^2 \leq C_{16} \beta_n \tag{2.3.16}$$

(Here $C_{16} = (C_5 \cdot C_{15})$

<u>Proof</u> As (2.1.4) and (2.1.10) hold and $EU_0 < \infty$, Theorem 2.3 also holds. This proves (2.3.44). (2.3.16) follows from (2.3.15) and Lemma 2.3.

CHAPTER 3

APPROXIMATIONS

3.1 Extensions

In this Section we extend the techniques of devising algorithms of a priori guaranteed learning properties. More distinctly, we derive conditions under which various modifications of the kernel-sequence W do not influence the asymptotic learning properties, and the behavior of the learning process may also be kept under control within finite training periods.

We obtain in this way possibilities to check the consequence of such approximations of implemental significance as (i) constraining the estimate to finite dimensions, (ii) round-offs and (iii) choosing, e.g., constant coefficient-spequences α and h , specifically, in terminated learning processes. In addition we also obtain some freedom in this way to embed, e.g., heuristic modifications into $W_n, n = 0,1,...$ in order to improve at early stages (i.e. in the "opening phase") learning.

We admit, as we did in Course I within the present Volume), only kernel-sequence W of the following properties: W is assumed to be, at any $n = 0,1,...$ related to some $\tilde{W}$ of a priori controlled learning properties in the following way:

$$W_n = \tilde{W}_n + \xi_n \tag{3.1.1}$$

Here $\tilde{\xi}_n$ is a random variable, taking values in $\mathcal{L}_2(\mu)$, such that

$$\sum_{n=0}^{\infty} \alpha_n^2 E\|\tilde{\xi}_n\|^2 < \infty \quad \text{and} \quad \sum_{n=0}^{\infty} \alpha_n E\|E(\tilde{\xi}_n|\eta^{n+1})\| < \infty \quad (3.1.2)$$

We assume that

$$E(\|\tilde{W}_n(X^n)\|^2|\eta^n) < C_{16}, \quad (3.1.3)$$

for all $n = 0,1,\ldots$ (We recall that $Z^n = \{Z_\nu, \nu \leq n\}$.)

We also assume for $\tilde{W}$ weak memory, in the sense that for some constraint $\tau_0 > 0$:

$$\|\tilde{W}_n(Z^n) - \tilde{W}_n(Z^{\overline{n-\nu,n}})\| \leq C_{17}\nu^{-\tau_0} \quad (3.1.4)$$

for all $n, \nu = 0,1,\ldots$ and any Z_n according to (3.1.3). $(Z^{\overline{n-\nu,n}}) = \{\tilde{Z}_n, n = 0,1,\ldots\}$, where $\tilde{Z}_i = Z_i$ for $n-\nu < i \leq n$, and $\tilde{Z}_i = 0$ for $i \leq n-\nu$.

We suppose that the increment of $\tilde{W}_n(X^{\overline{n-\nu,n}})$ may be linearly overbounded, provided we replace $X^{\overline{n-\nu,n}}$ by $_{X_n}Z^{\overline{n-\nu,n}}$ Viz.,

$$(3.1.5) \qquad |\tilde{\tilde{V}}_n(X^{\overline{n-\nu,n}}) - \tilde{\tilde{V}}_n({}_X Z^{\overline{n-\nu,n}})| \leq C_{18}\delta \qquad (3.1.5)$$

Here $_{X_n}Z^{\overline{n-\nu,n}} = \{\tilde{Z}_i, \ i \leq n\}$ where $\tilde{Z}_i = X_n$ if $n-\nu < i \leq n$, and $\tilde{Z} = 0$ if $i \leq n-\nu$, $\tilde{\tilde{V}}_n(Z^n) = \langle \tilde{X}_n, \tilde{W}_n(Z^n)\rangle$

(3.1.6) $$\delta = \sup_{n-\nu < i \le n} \| X_n - X_i \|$$

(Observe that we impose (3.1.5), concerning only certain perturbations of X^n, and not any Z^n.)

(3.1.7) $$V_n\left[\tilde{W}(Z^n)\right] = - E\left(\langle X_n - f, \tilde{W}_n(Z^n)\rangle \middle| \eta^n\right)$$

(Have f is the goal of the learning. E.g., $f = D^*, D, \Pi$. The minus in (3.1.7) is just for convenience. One gets by this choice e.g., for sequences, $\left\{W_n^{(4)}\right\}_{n=0}^{\infty} (i = 0,1), V_n\left(W^{(i)}(\chi^n)\right) \ge 0$.)

We do not assume that $\tilde{W}(X^n)$ exhibits at any n, some simple property which a priori assures for X_n, $n = 0,1,\ldots$ the wished learning properties. However, we confine ourselves to such W which are related to such "well behaved" kernel sequences, specifically, for $W({}_{xn}Z^n)$. Here ${}_{xn}Z^n = \{Z_i = X_n, i \le n\}$.)

For the sake of illustration, we take as such a reference sequence, specifically, $W^{(2)}(i=2)$. (According to (1.2.17)). We could equally well consider $W^{(0)}$, $W^{(1)}$(according to Examples 1.1 and 1.2 in Sec. 1.2) or any other kernel sequence, with a priori guaranteed learning properties.

More distinctly, we assume that $\tilde{W}$ and $W^{(2)}$are related as follows

(3.1.8) $$V_n\left[W^{(3)}({}_{xn}Z^n)\right] = V_n\left[\tilde{W}({}_{xn}Z^n)\right],$$

for any n.

In this more general case we have, however, to

tighten the constraints imposed on α, slightly. We, specifically, assume that in addition to (2.1.4a)

$$\sum_{n=0}^{\infty} \alpha_n n^{-\varepsilon} < \infty, \sum_{n=0}^{\infty} n^{2\tau} \left(\sup_{n-n^{\tau}<\nu\leq n} \alpha_\nu \right) < \infty \tag{3.1.9}$$

also holds, for any $\varepsilon < 0$ and some $0 < \tau < 1$. (Observe that (3.1.9) is met for any $\alpha_n = C_{19}(C_{20} + n)^{-1}$.)

<u>THEOREM 3.1</u> <u>If</u> (i) $\Pi \in H(K)$ and $E\|X_0\| < \infty$ (ii) P is such that, for any $\mathcal{F}(\eta^n)$-measurable random variable $Z_n \in H(K)$ we have $C_{13}\|Z_n\|^2 \leq E(Z_n(\omega_{n+1})^2 | \eta^n) \leq C_{14}\|Z_n\|^2$, (iii) $\{X_n\}_{n=0}^{\infty}$ is generated by (1.2.1) adopting a W, related to $W^{(2)}$ by the Conditions (3.1.1) through (3.1.5) and (3.1.8), ($W^{(2)}$ being such that $K(\omega,\omega) < C_{14} < \infty$) (iv) α is according to (2.1.4a) and (3.1.9) (v) Φ is uniformly norm reducing, with respect to $\mathcal{A}$ and (vi) we propose, at $n+1$, for classifying ω_{n+1} the decision rule (A_n, A_n^c) <u>then</u>

$$P\left(\lim_{n\to\infty} \|X_n - \Pi\| = 0\right) = 1 \tag{3.1.10}$$

$$P\left[\lim_{n\to\infty}\left(P_e(A_n) - P(\tilde{A})\right) = 0\right] = 1 \tag{3.1.11}$$

and also

$$P\left[\lim_{n\to\infty}\left(P_e(A_n,\eta^n) - P_e(\tilde{A},\eta^n)\right) = 0\right] = 1 . \tag{3.1.12}$$

The proof of Theorem 3.1 may be reduced to that of Theorem 2.8 by observing

LEMMA 3.1

$$E\left(\|X_{n+1}-\Pi\|^2 \mid \eta^n\right) \le \left(1+\zeta_n^{(4)}\right)\|X_n-\Pi\|^2 - 2\alpha_n V_n\left(W^{(2)}(x_n Z^n)\right) + \zeta_n^{(3)} \tag{3.1.13}$$

Here $\zeta_n^{(3)}$ is a non-negative valued $\mathfrak{F}(\eta^n)$-measurable random variables, for which $\sum_{n=0}^{\infty} E\zeta_n^{(3)} < \infty$. $\zeta_n^{(4)}$ is a non-negative constant such that $\sum_{n=0}^{\infty} \zeta_n^{(4)} < \infty$.

Remark Lemma 3.1 (which is a reproduction of Lemma 4.2 in Course I (within the present Volume) follows from (1.2.1) and the Conditions imposed on W and α.

More distinctly, from (1.2.1) and the uniformly norm reducing property of Φ we have

$$E\left(\|X_{n+1}-\Pi\|^2 \mid \eta^n\right) \le \|X_n-\Pi\|^2 + \\ + 2\alpha_n E\left(\langle X_n-\Pi, W_n(X^n)\rangle \mid \eta^n\right) + \alpha_n^2 E\left(\|W\|^2 \mid \eta^n\right) \tag{3.1.14}$$

$E\langle X_n-\Pi, W_n(X^n)\rangle \mid \eta^n)$ may be broken up into the sum of $E(\langle X_n-\Pi, \tilde{W}_n(X^n)\rangle \mid \eta^n)$ and a correction term.

We may, by Conditions (3.1.4) and (3.1.5), overbound

$$\left\|\tilde{W}_n(X^n) - \tilde{W}_n\left(X^{\overline{n-\nu,n}}\right)\right\|$$

$$\left| \tilde{\tilde{V}}_n\left(X^{\overline{n-\nu,n}}\right) - \tilde{\tilde{V}}_n\left({}_{x_n}Z^{\overline{n-\nu,n}}\right)\right|$$

$$\left\| \tilde{W}_n\left({}_{x_n}Z^{\overline{n-\nu,n}}\right) - \tilde{W}_n\left({}_{x_n}Z^{n}\right)\right\|$$

and replace, in this way $E\left(\langle X_n - \Pi, \tilde{W}_n(X_n)\rangle \middle| \eta^n\right)$ by the sum of $-V_n\left(\tilde{W}({}_{x_n}Z^n)\right)$, and a correction term. Contributions due to the correction terms may be included into either $\zeta_n^{(3)}$ or $\zeta_n^{(4)}$. Condition (3.1.3), the uniformly norm reducing property of Φ and the additional Conditions on W and α assure that $\sum_{n=0}^{\infty} E\,\zeta_n^{(3)} < \infty$. $\sum_{n=0}^{\infty} E\,\zeta_n^{(4)}$ follows from (3.1.2). Finally, by (3.1.8), we may replace $V_n\left[W({}_{x_n}Z^n)\right]$ by $V_n\left[W_n^{(2)}({}_{x_n}Z^n)\right]$. For further hints concerning the proof see Lemma 4.2 in Course I (within the present Volume).

One may in a similar way extend also Theorem 2.9 to any W, being related to $W^{(2)}$ by Conditions (3.1.1) through (3.1.5) and (3.1.8).

Viz., we have, for the behavior of such algorithms within finite training periods

THEOREM 3.2 Theorem 2.9 holds also for any W being related to $W^{(2)}$ by Conditions (3.1.1) through (3.1.5) and (3.1.8). Thus

$$EU_n < C_5\,\beta_n\,, \tag{3.1.15}$$

(3.1.16) $$E\left(P_e(A_n) - P_e(A)\right)^2 \le C_{19}\beta_n ,$$

and

(3.1.17) $$E\left(P_e(A_n,\eta^n) - P_e(A,\eta^n)\right)^2 \le C_{19}\beta_n .$$

Here

$$C_5 = \frac{EU_0\prod_{k=0}^{n_0}\left(1+\xi_k^{(4)}\right)+\sum_{k=0}^{n_0-1}E\xi_k^{(3)}\prod_{\ell=k+1}^{n_0}\left(1+\xi_\ell^{(4)}\right)+E\xi_{n_0}^{(3)}}{\beta_{n_0+1}}+\sum_{i=n_0}^{n}\frac{E\xi_i^{(4)}}{\beta_{i+1}}$$

(3.1.18)

$$\xi_i^{(3)} = \alpha_n\left(C_{23}n^{-\tau_1} + C_{20}n^{2\tau}\max_{n-n^{\tau}<i<n} + C_{21}\left\|E\left(\tilde{\xi}_n\,|\,\eta^{n+1}\right)\right\| + C_{22}\alpha_n\right).$$

(3.1.19)

<u>Remark</u> The expression for $\xi_i^{(3)}$ is obtained by e-valuating the additional correction terms included into $\xi_n^{(3)}$. The inspection of these constants is of interest as one may get in this way insight into how approximations of the kernels (e.g., the use of finite dimensional kernels, round offs, and various other appropriate modifications of the kernel W) influence the behavior of the learning process within finite training periods.

3.2 Finite Valued Estimates

Obviously the estimates X_n and the regulators $W_n (n = 0,1,\dots)$ taking values in some continuous space can not be precisely handled by digital computation. The algorithms become, however step by step implementable if (i) one constrains X_0 to some finite set B_0 , (ii) the values $\alpha_n W_n$ may take are, at any step n , members of some finite set $B_{n+1} (B_{n+1} \supset B_n)$ and (iii) Φ is such that $\Phi(Z) \in B_n$ for any $Z \in B_n$.

While the estimates $X_n (n = 0,1,\dots)$, devised in this way, may take at any step n just a finite member of admissible values, the iteration still may of course, as $n \to \infty$, run out of the actually available (or admissible) storage capacity, because of the increasing size of B_n .

Overflow of this sort is, of course, unnoticed if the iteration is termined before such overflow happens, by some stopping rule. Anyhow the significance of such problems is related to the behavior of the learning process within finite training periods.

We illustrate, in this Section, how may the asymptotic and finite time properties of iterative learning be studied if the estimates are, at any step, constrained to some finite range. (We take also in this case, as a reference, $W^{(2)}$ just as an illustration.)

THEOREM 3.3 If (i) $\Pi \in \mathcal{A} \subset H(K)$, $\mathcal{A}$ being a bounded set and $X_0 \in B_0$ where $\{B_n\}_{n=0}^{\infty}$ is a sequence of finite sets such that $B_{n+1} \supset B_n$ and $\lim_{n\to\infty} \|Z - B_n\| = 0$ for any $Z \in A$, (ii) P is such that for any $\mathcal{F}(\eta^n)$-measurable random variable $Z_n \in H(K)$, we have $C_{12}\|Z_n\|^2 \leq E(Z_n(\omega_{n+1})^2|\eta^n) \leq C_{14}\|Z_n\|^2$, (iii) we generate finite valued estimates $\{X_n\}_{n=0}^{\infty}$ according to (1.2.1), adopting a W for which $\alpha_n W_n \in B_{n+1}$ (iv) assume that W is related to $W^{(2)}$ by Conditions (3.1.1) through (3.1.5) and (3.1.8) (where $W^{(2)}$ is such that $K(\omega,\omega) < C_{14} < \infty$) (v) α is according to (2.2.4a) and (3.1.9) (vi) Φ is uniformly norm reducing with respect to $\mathcal{A}$ and (vii) we propose, at $n+1$, for classifying ω_{n+1} the decision rule (A_n, A_n^c), then

(3.2.1) $$P\left(\lim_{n\to\infty} \|X_n - \Pi\| = 0\right) = 1$$

(3.2.2) $$P\left(\lim_{n\to\infty} \left(P_e(A_n) - P(A)\right) = 0\right) = 1$$

and also

(3.2.3) $$P\left[\lim_{n\to\infty} \left(P_e(A_n,\eta^n) - P_e(A,\eta^n\right) = 0\right] = 1$$

(Here

(3.2.4) $$\|Z - B_n\| = \inf_{\tilde{Z}\in B_n} \|Z - \tilde{Z}\| \text{ and } W_n = W(X^n).)$$

Remark As we merely tightened the conditions imposed on W by the constraint $\alpha W_n \in B_{n+1}$, Theorem 3.3 immediately follows from Theorem 3.1. In the same way Theorem 3.2 may also be interpreted for finite valued statistics, generated in the aforementioned sense.

Let us now briefly consider the question how may such a W be found actually. One may arrive at an appropriate choice of W by adopting series truncation and roundoff afteranother.

Let $\left\{ \varphi_i \right\}_{i=\infty}^{\infty}$ be a complete orthonormal set within $H(K)$. Assume a $H(K)$, for which $W_n^{(2)} = \sum_{i=1}^{\infty} C_i^{(n)} \varphi_i$, and let $W_n^{(A)} = \sum_{i=1}^{N} C_i^{(n)} \varphi_i$ be some N -dimensional approximation of $W_n^{(2)}$ / $N < \infty$./ The series truncation error is, by definition, $\tilde{\xi}_n^{(A)} = W_n^{(A)} - W_n^{(2)}$.

Let, at any n , B_n be a finite set defined by a system of equidistant rectangular cells, spanned by $\left\{ \varphi_i \right\}_{i=1}^{N}$ and covering $\mathcal{A}$ within accuracy of the cell-size. Let the cell-size (i.e. the size of the roundoff steps), specifically along the i -th coordinate axis, be $\Delta_n^{(i)}$. Let $\alpha_n W_n$ be the rounded version of the kernel $\alpha_n W_n^{(A)}$ and $\tilde{\xi}_n^{(B)} = W_n - W_n^{(A)}$ dende the roundoff error introduced, with respect to $W_n^{(A)}$, in this way. Let $\Delta_n = \left(\sum_{i=1}^{N} \Delta_n^{(i)^2} \right)^{1/2}$. As $\alpha_n W_n \in B_{n+1}$, we obviously have $\alpha \| \tilde{\xi}_n^{(B)} \| < \Delta_{n+1}$.

A simple way to meet (3.1.2) is to have $\| \tilde{\xi}_n^{(A)} \| < C_{91}, \| \tilde{\xi}_n^{(B)} \| < C_{92}$, $\sum_{n=\infty}^{\infty} \alpha_n \| \tilde{\xi}_n^{(A)} \| < \infty$ and $\sum_{n=0}^{\infty} \alpha_n E \| E(\tilde{\xi}_n^{(B)} | \eta^n) \| < \infty$.(We will consider this constraint, concerning specifically round-off, in the sequel.)

It still needs some explanation how to meet the condition

$$\Phi(Z) \in B_{n+1} \tag{3.2.5}$$

for any $Z \in B_{n+1}$.

Let us define $\mathcal{A}$ coordinatewise and Φ by $\left\{ \Phi \right\}_{i=1}^{N}$,
according to (1.2.14). Obviously (3.2.5) is met if, for all ,
a_i and b_i are rounded values along the i-th coordinate axis.

Let us confine ourselves, within the rest of this section, to errors due to roundoff. In order to specify roundoff rules concisely, we introduce the following notation:

$$\left\lfloor <\alpha_n W_n^{(A)}, \varphi_i> \right\rfloor = \Delta_n^{(i)} \text{ ent} \left(<\alpha_n W_n^{(A)}, \varphi_i> / \Delta_n^{(i)} \right)$$

and

$$\left\lceil <\alpha_n W_n^{(A)}, \varphi_i> \right\rceil = \left\lfloor <\alpha_n W_n^{(A)}, \varphi_i> \right\rfloor + \Delta_n^{(i)}$$

for $i = \overline{1, N}$

Let us consider two sort of round-off rules, called SIMPLE and CONDITIONALLY UNBIASED round-off in the sequel. For simple roundoff:

$$<\alpha_n W_n, \varphi_i> = \left\lfloor \alpha_n W_n^{(A)}, \varphi_i> \right\rfloor$$

if

$$\left\lceil <\alpha_n W_n^{(A)}, \varphi_i> \right\rceil - <\alpha_n W_n^{(A)}, \varphi_i> \geq$$

$$\geq <\alpha_n W_n^{(A)}, \varphi_i> - \left\lfloor \alpha_n W_n^{(A)}, \varphi_i> \right\rfloor$$

and

$$\langle \alpha_n W_n , \varphi_i \rangle = \left[\alpha_n W_n^{(A)} , \varphi_i \rangle \right]$$

otherwise,

First, let us consider the constraints (3.1.2) imposed on the error $\tilde{\xi}_n$ (including among other contributions also the roundoff error). From $\alpha_n \| \tilde{\xi}_n^{(B)} \| \leq \Delta_{n+1}$ follows

$$\sum_{n=\infty}^{\infty} \alpha_n E\left[E\left(\| \tilde{\xi}_n^{(B)} \| \, \Big| \, \eta^{n+1} \right) \right] < \sum_{n=0}^{\infty} \Delta_{n+1}$$

Obviously, for simple round-off, both constraints in (3.1.2) are met if $\left\{ \Delta_n \right\}_{n=0}^{\infty}$ is such that

$$\sum_{n=0}^{\infty} \Delta_{n+1} < \infty .$$

(Observe that we arrived at this condition by overbounding what may happen in the course of subsequent roundoffs as a worst case. It is likely, however, that round-off errors partly compensate each other.)

It is well known (Fabian, 1957 and Schmetterer 1970) that one may entirely remove the round off errors, related to the second condition in (3.1.2), at the expense of additional complexity, by using rounded values which are conditionally unbiased estimates of the true values. Viz.,

$$E\left(\alpha_n W_n \middle| \eta^{n+1}\right) = \alpha_n W_n^{(A)}$$

This condition is met if one adopts a randomized round off rule, the randomization of which is carried out, for each coordinate $\langle \alpha_n W_n^{(A)}, \varphi_i \rangle$, totally independently in the following way: The probability of having

$$\langle \alpha_n W_n, \varphi_i \rangle = \lfloor \langle \alpha_n W_n^{(A)}, \varphi_i \rangle \rfloor$$

is

$$\lceil \langle \alpha_n W_n^{(A)}, \varphi_i \rangle \rceil - \langle \alpha_n W_n^{(A)}, \varphi_i \rangle .$$

In this case of a conditionally unbiased roundoff the second constraint in (3.1.2) is necessarily met. Observe that, because of $\alpha_n \| \tilde{\xi}_n^{(B)} \| < \Delta_{n+1}$,

$$\sum_{n=0}^{\infty} \alpha_n^2 \| \xi_n^{(B)} \|^2 < \sum_{n=0}^{\infty} \Delta_{n+1}^2 .$$

Thus, specifically for conditionally unbiased roundoff, the first constraint in (3.1.2) is met by any $\{\Delta_n\}_{n=0}^{\infty}$, for which

$$\sum_{n=0} \Delta_{n+1}^2 < \infty .$$

The practical question is, of course, how round

off errors actually influence the behavior of the learning process within finite training periods. One may get some insight also in this respect by concerning the contribution which appears, due to $\tilde{\xi}_n$, in (3.1.19) and (2.1.14). We do not, however, enter into a further discussion of this topic within the present context.

REFERENCES

Aĭzerman, M.A.; Braverman, E.M.; Rozonoer, L.I., "Teoretičeskie osnovy metoda potencial'nyh funkcii v zadače ob obučenii avtomatov razdelniju vhodnyh situacii na klassy" Avtomatika i Telemehanika. No. 9., pp. 917-936, 1964a.

Aĭzerman, M.A.; Braverman, E.M.; Rozonoer, L.I., "Verojatnosnaja zadača ob obučenii avtomatov raspoznavaniju klassov i metod potencialnyh funkcii" Avtomatika i Telemehanika. No. 6, pp. 1307-1323, 1964b.

Aĭzerman, M.A.; Braverman, E.M.; Rozonoer, L.I., "Method potencial'nyh funkcii v teorii obučenija mašin". Nauka, 1970.

Aronszajn, N., "Theory of reproducing kernels" Trans. Amer., Math. Soc., Vol. 63,pp. 337-404, 1950.

Braverman, E.M.; Pjatnickij, E.S., "Ocenka skorosti shodimosti algoritmov, osnovannyh na metode potenicalnyh funkcii". Avtomatika i Telemehanika. No. 1, pp. 95-112, 1966.

Braverman, E.M.; Rozonoer, L.I., "Shodimost'slucainyh processov v teorii obučenija masin I." Avtomatika i Telemehanika No. 1., pp. 57-77, 1969a.

Braverman, E.M.; Rozonoer, L.I., "Shodimost'slučainyh processov v teorii obučenija masin II." Avtomatika i Telemehanika, pp. 87-103, 1969b.

Cover, T.M. "Learning in pattern recognition" in Watanabe, M.S. (ed.) "Methodologies of Pattern Recognition" Academic Press, pp. 111-113, 1969b.

Csibi S., "On iteration rules with memory in machine learning" Problems of Control and Information Theory. Vol. 1., No. 1. pp. 37-50, 1972, partly appeared in the Proceedings of the Fourth Hawaii International Conference on System Sciences, Western Periodicals, pp. 205-207, 1971a.

Csibi, S., "On embedding heuristics and including complexity constraints into convergent learning algorithms" in Watanabe, M.S. (ed.) "Frontiers of Pattern Recognition", Academic Press, 1972

Csibi, S., "Approximation in learning decision functions recursively, Trans. 6th Prague Conf. on Info. Theory Statistical Decision Functions, Random Processes, Czech. Acad. Sci. Prague 1973e

Csibi, S., "Statistical learning processes", Preprint, Telec. Res. Inst., Budapest, 1973b (in Hungarian)

Csibi, S., "Learning finite-dimensional projections of decision functions from weakly dependent samples", Preprint Telec. Res. Inst., Budapest, 1973c

Cypkin, Ja. Z., "Adaptacija i obučenie v avtomatičeskih sistemah" Nauka, 1968.

Cypkin, J.Z., "Osnovy teorii obučajuščihsja sistem" Nauka, 1970.

Cypkin, J.Z., Kel'mans, G.K., "Adaptivnyi baesov podhod" Problemy peredači Informacii, Vol. 10, No. 1., 1970.

Fabian, V., Zufälliges Abrunden und die Konvergenz des linearen (Seidelschen) Iterationsverfahrens", Math.Nachr. 16, pp. 256-259, 1957.

Fu, K.S., "Sequential methods in pattern recognition and machine learning" Academic Press, 1968.

Fu, K.S., "Learning control systems - Review and outlook" IEE Trans. Automatic Control, Vol. AC-15, No. 2, pp. 210-221, 1970.

Gulyás, O., "On extended potential function type learning algorithms and their convergence rate" <u>Problems of Control and Information Theory</u>, Vol.1, No.1 1972

Győrfi, L., "Estimation of probability density and optimal decision function in RKHS" <u>Proc. of the European Meeting of Statisticians</u>, Budapest, 1972

Győrfi, L., "On the estimation of asymptotic error probability", (Corr.) IEEE Trans. IT-20, March 1974 (to appear)

Kailath, T. " Some results on singular detection" Tech. Rept. No. 7050-2, <u>Stanford El. Lab., Stanford Univ</u>., 1965.

Meschkowski, H., "Hilbersche Räume mit Kernfunktion", Springer, 1962

Parzen, E., "On estimation of a probability density function and mode" <u>Annals. Math. Statist</u>., Vol. 33, pp. 1065-1076, 1962.

Révész, P.; Rejto Lidia, "Density estimation and pattern classification", <u>Problems of Control and Information</u> Theory, 1972

Rozanov, Ju., A., "<u>Stacionarnye slučainye processy</u>" FM, 1963.

Rozenblatt, M. "A central limit theorem and a strong mixing condition" Proc. Nat. Acad. Sci., 42, 1, 1956.

Schmetterer, L., "Stochastic approximation" <u>Proc. 4th Berkeley Symposium on Math. Statist. and Probability, Univ. Calif</u>. Press, pp. 587-608, 1960.

Saridis, G.N. "Learning applied to successive approximation algorithms" <u>IEEE Trans. System Science and Cybernetics</u>. Vol. SSC-6, No. 2, pp. 97-103, 1970.

Watanabe, M.S. (ed.) "<u>Methodologies of Pattern Recognition</u>", Academic Press, 1969.

Watanabe, M.S., "Frontiers of Pattern Recognition", Academic Press, 1973.

Wolverton, Ch.T.; Wagner, T.J., "Asymptotically optimal discriminant functions for pattern recognition", IEEE Trans. Information Theory, Vol. IT-15, No. 2, pp. 258-265, 1969.

CONTENTS

Printed in the United States
By Bookmasters